CAMBRIDGE TRACTS IN MATHEMATICS

General Editors

B. BOLLOBÁS, W. FULTON, F. KIRWAN,
P. SARNAK, B. SIMON, B. TOTARO

**220   Lectures on Contact 3-Manifolds, Holomorphic Curves
and Intersection Theory**

CAMBRIDGE TRACTS IN MATHEMATICS

GENERAL EDITORS
B. BOLLOBÁS, W. FULTON, F. KIRWAN,
P. SARNAK, B. SIMON, B. TOTARO

A complete list of books in the series can be found at www.cambridge.org/mathematics.
Recent titles include the following:

186. Dimensions, Embeddings, and Attractors. By J. C. ROBINSON
187. Convexity: An Analytic Viewpoint. By B. SIMON
188. Modern Approaches to the Invariant Subspace Problem. By I. CHALENDAR
    and J. R. PARTINGTON
189. Nonlinear Perron–Frobenius Theory. By B. LEMMENS and R. NUSSBAUM
190. Jordan Structures in Geometry and Analysis. By C.-H. CHU
191. Malliavin Calculus for Lévy Processes and Infinite-Dimensional Brownian Motion.
    By H. OSSWALD
192. Normal Approximations with Malliavin Calculus. By I. NOURDIN and G. PECCATI
193. Distribution Modulo One and Diophantine Approximation. By Y. BUGEAUD
194. Mathematics of Two-Dimensional Turbulence. By S. KUKSIN and A. SHIRIKYAN
195. A Universal Construction for Groups Acting Freely on Real Trees.
    By I. CHISWELL and T. MÜLLER
196. The Theory of Hardy's Z-Function. By A. IVIĆ
197. Induced Representations of Locally Compact Groups. By E. KANIUTH and K. F. TAYLOR
198. Topics in Critical Point Theory. By K. PERERA and M. SCHECHTER
199. Combinatorics of Minuscule Representations. By R. M. GREEN
200. Singularities of the Minimal Model Program. By J. KOLLÁR
201. Coherence in Three-Dimensional Category Theory. By N. GURSKI
202. Canonical Ramsey Theory on Polish Spaces. By V. KANOVEI, M. SABOK,
    and J. ZAPLETAL
203. A Primer on the Dirichlet Space. By O. EL-FALLAH, K. KELLAY, J. MASHREGHI,
    and T. RANSFORD
204. Group Cohomology and Algebraic Cycles. By B. TOTARO
205. Ridge Functions. By A. PINKUS
206. Probability on Real Lie Algebras. By U. FRANZ and N. PRIVAULT
207. Auxiliary Polynomials in Number Theory. By D. MASSER
208. Representations of Elementary Abelian p-Groups and Vector Bundles. By D. J. BENSON
209. Non-homogeneous Random Walks. By M. MENSHIKOV, S. POPOV, and A. WADE
210. Fourier Integrals in Classical Analysis (Second Edition). By C. D. SOGGE
211. Eigenvalues, Multiplicities and Graphs. By C. R. JOHNSON and C. M. SAIAGO
212. Applications of Diophantine Approximation to Integral Points and Transcendence.
    By P. CORVAJA and U. ZANNIER
213. Variations on a Theme of Borel. By S. WEINBERGER
214. The Mathieu Groups. By A. A. IVANOV
215. Slenderness I: Foundations. By R. DIMITRIC
216. Justification Logic. By S. ARTEMOV and M. FITTING
217. Defocusing Nonlinear Schrödinger Equations. By B. DODSON
218. The Random Matrix Theory of the Classical Compact Groups. By E. S. MECKES
219. Operator Analysis. By J. AGLER, J. E. MCCARTHY, and N. J. YOUNG
220. Lectures on Contact 3-Manifolds, Holomorphic Curves and Intersection Theory.
    By C. WENDL

Lectures on Contact 3-Manifolds, Holomorphic Curves and Intersection Theory

CHRIS WENDL
Humboldt University of Berlin

CAMBRIDGE
UNIVERSITY PRESS

University Printing House, Cambridge CB2 8BS, United Kingdom

One Liberty Plaza, 20th Floor, New York, NY 10006, USA

477 Williamstown Road, Port Melbourne, VIC 3207, Australia

314–321, 3rd Floor, Plot 3, Splendor Forum, Jasola District Centre,
New Delhi – 110025, India

79 Anson Road, #06-04/06, Singapore 079906

Cambridge University Press is part of the University of Cambridge.

It furthers the University's mission by disseminating knowledge in the pursuit of
education, learning, and research at the highest international levels of excellence.

www.cambridge.org
Information on this title: www.cambridge.org/9781108497404
DOI: 10.1017/9781108608954

© Chris Wendl 2020

This publication is in copyright. Subject to statutory exception
and to the provisions of relevant collective licensing agreements,
no reproduction of any part may take place without the written
permission of Cambridge University Press.

First published 2020

A catalogue record for this publication is available from the British Library.

ISBN 978-1-108-49740-4 Hardback

Cambridge University Press has no responsibility for the persistence or accuracy
of URLs for external or third-party internet websites referred to in this publication
and does not guarantee that any content on such websites is, or will remain,
accurate or appropriate.

Contents

Preface		*page* vii
Acknowledgments		ix

Introduction: Motivation		1
Lecture 1	**Closed Holomorphic Curves in Symplectic 4-Manifolds**	11
1.1	Some Examples of Symplectic 4-Manifolds	11
1.2	McDuff's Characterization of Symplectic Ruled Surfaces	17
1.3	Local Foliations by Holomorphic Spheres	22
Lecture 2	**Intersections, Ruled Surfaces, and Contact Boundaries**	26
2.1	Positivity of Intersections and the Adjunction Formula	26
2.2	Application to Ruled Surfaces	32
2.3	Contact Manifolds, Symplectic Fillings and Cobordisms	35
2.4	Asymptotically Cylindrical Holomorphic Curves	38
Lecture 3	**Asymptotics of Punctured Holomorphic Curves**	44
3.1	Holomorphic Half-Cylinders as Gradient-Flow Lines	45
3.2	Asymptotic Formulas for Cylindrical Ends	49
3.3	Winding of Asymptotic Eigenfunctions	53
3.4	Local Foliations and the Normal Chern Number	55
Lecture 4	**Intersection Theory for Punctured Holomorphic Curves**	62
4.1	Statement of the Main Results	62
4.2	Relative Intersection Numbers and the $*$-Pairing	66
4.3	Adjunction Formulas, Relative and Absolute	71
Lecture 5	**Symplectic Fillings of Planar Contact 3-Manifolds**	77
5.1	Open Books and Lefschetz Fibrations	77
5.2	A Classification Theorem for Symplectic Fillings	83
5.3	Sketch of the Proof	86

Appendix A	**Properties of Pseudoholomorphic Curves**	94
A.1	The Closed Case	94
A.2	Curves with Punctures	102
Appendix B	**Local Positivity of Intersections**	108
B.1	Regularity and the Similarity Principle	108
B.1.1	Linear Cauchy–Riemann Type Operators	109
B.1.2	Elliptic Regularity	111
B.1.3	Local Existence of Holomorphic Sections	120
B.1.4	The Similarity Principle	123
B.2	The Representation Formula	126
B.2.1	The Generalized Tangent-Normal Decomposition	128
B.2.2	A Lemma on Normal Push-Offs	130
B.2.3	Local Coordinates	134
B.2.4	Constructing the Normal Push-Off	137
B.2.5	Conclusion of the Proof	146
B.3	Counting Local Intersections and Singularities	148
Appendix C	**A Quick Survey of Siefring's Intersection Theory**	158
C.1	Preliminaries	158
C.2	The Intersection Pairing	160
C.3	The Adjunction Formula	163
C.4	Covering Relations	166
C.5	The Intersection Product of Buildings	168
C.6	Comparison with the ECH Literature	174
References		177
Index		182

Preface

The main portion of this book is a lightly revised set of expository lecture notes written originally for a five-hour minicourse on the intersection theory of punctured holomorphic curves and its applications in 3-dimensional contact topology, which I gave as part of the LMS Short Course "Topology in Low Dimensions" at Durham University, August 26–30, 2013. These lectures were aimed primarily at students, and they required only a minimal background in holomorphic curve theory since the emphasis was on topological rather than analytical issues. The original appendices were relatively brief, their purpose being to provide a quick survey of analytical background material on holomorphic curves that I needed to refer to in the lectures without assuming that students already knew it. In revising the manuscript for publication, I have added a motivational introduction, and taken the opportunity to insert two further additions that I felt were lacking from the existing literature, as a result of which the appendices have become considerably more substantial. One (Appendix B) is a complete proof of local positivity of intersections, including just enough background material on elliptic regularity for a student familiar with distributions and Sobolev spaces to consider it "self-contained"; this notably includes a weak version of the Micallef–White theorem, which some readers may, I hope, find easier to comprehend than the deeper result in [MW95] that inspired it. The other (Appendix C) is a quick survey of Siefring's intersection theory of punctured holomorphic curves, putting the essential facts and formulas in as compact a form as possible for the benefit of researchers who need a ready reference. Most of what is in Appendix C also appears in Lectures 3 and 4, but the latter are written in a more pedagogical style that develops the structure of the theory based on a few core ideas – which is presumably helpful if your goal is to understand why the main results are true, but less so if you just need to look up a specific formula, and Appendix C is there to help in that case.

viii   *Preface*

Intersection theory has played a prominent role in the study of closed symplectic 4-manifolds since Gromov's paper [Gro85] on pseudoholomorphic curves, leading to a myriad of beautiful rigidity results that are either not accessible or not true in higher dimensions. In the last 15 years, the highly nontrivial extension of this theory to the case of punctured holomorphic curves, due to Siefring [Sie08, Sie11], has led to similarly beautiful results about contact 3-manifolds and their symplectic fillings. These lectures begin with an overview of the closed case and an easy application (McDuff's characterization of symplectic ruled surfaces), and then explain the essentials of Siefring's intersection theory and how to use it in the real world. As a sample application, Lecture 5 concludes by discussing the classification of symplectic fillings of planar contact manifolds via Lefschetz fibrations [Wen10b].

How to use these lecture notes: I expect a variety of audiences to find these lecture notes useful for a variety of reasons. Since they were written with an audience of students in mind, I did not want to assume too much previous knowledge of symplectic/contact geometry or holomorphic curves, and most of the text reflects that. On the other hand, I also expect a certain number of readers to be experienced researchers who already know the essentials of holomorphic curve theory – including the adjunction formula in the closed case – but would specifically like to learn about the intersection theory for *punctured* curves. For readers in this category, I recommend starting with Appendix C for an overview of the basic facts, and then turning back to Lectures 3 and 4 for details whenever necessary. If, on the other hand, you are a student and still getting to know the field of symplectic and contact topology, you'll probably wish to start from the beginning.

Or if you really want to challenge yourself, feel free to read the whole thing backward.

Acknowledgments

I would like to thank Richard Siefring and Michael Hutchings for many conversations over the years that have improved my understanding of the subjects discussed in this book. Thanks are also due to Andrew Lobb, Durham University and the London Mathematical Society for bringing about the summer school that gave rise to the original notes. They were written mostly while I worked at University College London, with partial support from a Royal Society University Research Fellowship and a Leverhulme Trust Research Project Grant.

Introduction

Motivation

In order to illustrate briefly what these lectures are about, I'd like to give an informal sketch of two closely related theorems from the early days of symplectic topology. The first is a beautiful application of the theory of closed pseudoholomorphic curves as introduced by Gromov in [Gro85], and its proof requires only a few basic facts from this theory, plus some knowledge of the standard homological intersection product from algebraic topology. The second theorem admits a closely analogous proof, but we will see that the intersection-theoretic portion of the argument is difficult to make precise, because it is no longer homological – it requires some generalization of the intersection product in which "cycles" need not be closed. One of the main objectives of the subsequent lectures will be to make this idea precise and demonstrate what else it can be used for.

The statements of these theorems assume familiarity with the notions of minimal symplectic 4-manifolds, symplectomorphisms, symplectic submanifolds, the standard symplectic structure on $\mathbb{R}^4$, the sign of a transverse intersection, and the homological intersection product – some background on all of these topics is covered in Lectures 1 and 2.

Theorem 1 *Suppose (M, ω) is a closed, connected, minimal symplectic 4-manifold containing a pair of symplectic submanifolds $S_1, S_2 \subset M$ with the following properties:*

- *Both are homeomorphic to S^2.*
- *Both have vanishing homological self-intersection number:*

$$[S_1] \cdot [S_1] = [S_2] \cdot [S_2] = 0.$$

- *The set $S_1 \cap S_2 \subset M$ consists of a single transverse and positive intersection.*

2 *Motivation*

Then there exists a symplectomorphism identifying (M, ω) with $(S^2 \times S^2, \omega_0)$ such that S_1 and S_2 are identified with $S^2 \times \{const\}$ and $\{const\} \times S^2$, respectively, and ω_0 is a product of two area forms on S^2.

This result says in effect that if we are given a certain type of "local" information about submanifolds of a closed symplectic 4-manifold, then this is enough to recover its global structure. From an alternative perspective, it says that the vast majority of closed symplectic 4-manifolds do not contain certain types of symplectic submanifolds. The second result says something similar, but now the symplectic manifold is noncompact and the "local" information we are given is its structure outside of some compact subset – the theorem is typically summarized by saying that there do not exist any exotic symplectic 4-manifolds that look "standard at infinity."

Theorem 2 *Suppose (M, ω) is an open, connected, minimal symplectic 4-manifold with a compact subset $K \subset M$ such that $(M \backslash K, \omega)$ is symplecto-morphic to the complement of a compact subset in the standard symplectic $\mathbb{R}^4$. Then (M, ω) is globally symplectomorphic to the standard symplectic $\mathbb{R}^4$.*

Remark 3 Both of these theorems appeared in less general forms in Gromov's paper [Gro85] (see §2.4.A'_1 and §3.C, respectively). The statements given above are attributed to both Gromov and McDuff, as they rely on the slightly more sophisticated intersection theory of closed holomorphic curves that was developed by McDuff within a few years after Gromov's paper (see, in particular, [McD90]). Theorem 2 can also be rephrased as the statement that S^3 with its standard contact structure admits a unique minimal symplectic filling, and we will discuss this version of the result in Lecture 5 (see, in particular, Corollary 5.7).

Let's sketch a proof of Theorem 1. The starting point is the observation that since S_1 and S_2 are both *symplectic* submanifolds and their intersection is transverse and positive, one can choose a compatible almost complex structure $J \colon TM \to TM$ on (M, ω) that preserves the tangent spaces of S_1 and S_2 (see §1.1 for more on almost complex structures). This makes S_1 and S_2 into images of embedded *J-holomorphic spheres*, i.e., smooth maps $u \colon S^2 \to M$ that satisfy the *nonlinear Cauchy–Riemann equation*

$$Tu \circ i = J \circ Tu,$$

where $i \colon TS^2 \to TS^2$ is the almost complex structure on S^2 resulting from its standard identification with the extended complex plane $\mathbb{C} \cup \{\infty\}$. The advantage of replacing symplectic submanifolds by J-holomorphic spheres is a matter of rigidity: the condition of being a symplectic submanifold is open and

Motivation 3

thus quite flexible, i.e., the space of all symplectic submanifolds is unmanageably large, whereas J-holomorphic spheres are solutions to an elliptic PDE, and thus tend to come in finite-dimensional moduli spaces, which are sometimes (if we're lucky!) even compact. For this reason, we now consider for each $k = 1, 2$ the *moduli spaces*

$$\mathcal{M}_k(J) := \left\{ u \colon S^2 \to M \mid Tu \circ i = J \circ Tu \text{ and } [u] := u_*[S^2] \right.$$
$$\left. = [S_k] \in H_2(M) \right\} \Big/ \operatorname{Aut}(S^2, i),$$

where $\operatorname{Aut}(S^2, i)$ is the group of holomorphic automorphisms $\varphi \colon S^2 \to S^2$ of the extended complex plane (i.e., the Möbius transformations), acting on the space of J-holomorphic maps $u \colon S^2 \to M$ by $\varphi \cdot u := u \circ \varphi$. We assign to this space the natural topology arising from C^∞-convergence of maps. Both $\mathcal{M}_1(J)$ and $\mathcal{M}_2(J)$ are clearly nonempty, since they contain equivalence classes of parametrizations of the submanifolds S_1 and S_2, respectively. One can now apply general results from the theory of J-holomorphic curves to prove that for generic choices of the almost complex structure J, $\mathcal{M}_1(J)$ and $\mathcal{M}_2(J)$ both are compact smooth 2-dimensional manifolds. A quick survey of the analytical results behind this is given in Appendix A.1, and we will sketch the proof in a somewhat more general setting in Lectures 1 (see Lemmas 1.17 and 1.18) and 2, though we do not plan to go too deeply into such analytical details in this book.

What we will discuss in more detail is the intersection-theoretic properties of the J-holomorphic spheres in $\mathcal{M}_1(J)$ and $\mathcal{M}_2(J)$. We observe first that the hypotheses of Theorem 1 clearly imply

$$[S_1] \cdot [S_2] = 1,$$

as this intersection number can be computed as a signed count of transverse intersections between S_1 and S_2, for which there is only one intersection to count, and it is positive. In Lecture 2 and Appendix B, we will discuss a standard result known as *positivity of intersections*, which implies that whenever $u \colon \Sigma \to M$ and $v \colon \Sigma' \to M$ are two closed J-holomorphic curves with nonidentical images in an almost complex 4-manifold M, their intersections are all isolated and count positively toward the homological intersection number $[u] \cdot [v] \in \mathbb{Z}$; moreover, the contribution of each isolated intersection is exactly $+1$ if and only if that intersection is transverse. This is very strong information, from which one can deduce the following:

(1) For each $k = 1, 2$ and every pair of distinct elements $u, v \in \mathcal{M}_k(J)$, the images of $u \colon S^2 \to M$ and $v \colon S^2 \to M$ are disjoint. (This follows from the condition $[S_k] \cdot [S_k] = 0$.)

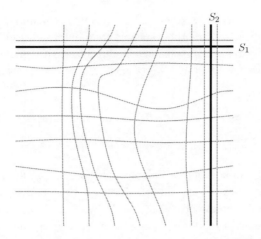

Figure 1 The two symplectic submanifolds $S_1, S_2 \subset M$ generate two transverse foliations by holomorphic spheres in the proof of Theorem 1. The two families can be regarded as a "coordinate grid" that identifies M with $S^2 \times S^2$.

(2) For every $u \in \mathcal{M}_1(J)$ and $v \in \mathcal{M}_2(J)$, the maps $u \colon S^2 \to M$ and $v \colon S^2 \to M$ have exactly one intersection point, which is transverse and positive.

A related result discussed in §2.1, called the *adjunction formula*, makes it possible to characterize in homological terms which J-holomorphic curves in an almost complex 4-manifold are embedded, and in this case it implies

(3) Every element of $\mathcal{M}_1(J)$ or $\mathcal{M}_2(J)$ is embedded.

Finally, we will see in §1.3 that whenever $u \in \mathcal{M}_k(J)$ is an embedded J-holomorphic sphere in one of these moduli spaces, the 2-parameter family of nearby J-holomorphic spheres in $\mathcal{M}_k(J)$ forms a smooth foliation of the neighborhood of $u(S^2)$ in M. Combining this with the compactness of $\mathcal{M}_k(J)$, it follows that the set of points in M that are contained in the images of any of the spheres in $\mathcal{M}_k(J)$ is both open and closed, and thus it is everything: the holomorphic spheres of $\mathcal{M}_k(J)$ foliate M. The result is the "coordinate grid" depicted in Figure 1: starting from the two symplectically embedded spheres $S_1, S_2 \subset M$, we obtain two smooth families of embedded J-holomorphic spheres that each foliate M such that each sphere in $\mathcal{M}_1(J)$ has a unique transverse intersection with each sphere in $\mathcal{M}_2(J)$. It follows that there is a diffeomorphism

$$M \xrightarrow{\cong} \mathcal{M}_1(J) \times \mathcal{M}_2(J), \tag{1}$$

assigning to each point $p \in M$ the unique pair of holomorphic spheres $(u, v) \in \mathcal{M}_1(J) \times \mathcal{M}_2(J)$ such that both have p in their images. Moreover, for each individual element of $\mathcal{M}_1(J)$ parametrized by a map $u : S^2 \to M$, there is a diffeomorphism

$$S^2 \xrightarrow{\cong} \mathcal{M}_2(J)$$

sending each $z \in S^2$ to the unique holomorphic sphere $v \in \mathcal{M}_2(J)$ that has $u(z)$ in its image; this proves that $\mathcal{M}_2(J)$ has the topology of S^2, and, in the same manner, one shows $\mathcal{M}_1(J) \cong S^2$. In summary, (1) can now be interpreted as a diffeomorphism from M to $S^2 \times S^2$. There is still a bit of work to be done in identifying the symplectic structure ω with a product of two area forms, but the techniques needed for this are not hard – they involve geometric tools such as the Moser stability theorem for deformations of symplectic forms (see, e.g., [MS17]), but no serious analysis is required.

The original proof of Theorem 2 used a clever "capping" trick to derive it from Theorem 1. For this motivational discussion, I would like to sketch a different proof that is conceptually simpler, but trickier in the technical details.

By the hypotheses of Theorem 2, we can decompose the open symplectic manifold (M, ω) into two regions: one is the compact (but otherwise completely unknown) subset $K \subset M$, and the other is a region that we can identify with $(\mathbb{R}^4 \backslash K', \omega_{\mathrm{std}})$ for some compact set $K' \subset \mathbb{R}^4$, where ω_{std} denotes the standard symplectic form on $\mathbb{R}^4$. We would like to argue as we did in Theorem 1; that is, find a nice pair of "seed curves" to generate two well-behaved moduli spaces of J-holomorphic curves that can then be used to form a coordinate grid identifying M with $\mathbb{R}^4$. One easy way to find such seed curves is by observing that $\mathbb{R}^4$ has a natural identification with $\mathbb{C}^2$ such that the natural multiplication by i on $\mathbb{C}^2$ defines a compatible almost complex structure on $(\mathbb{R}^4, \omega_{\mathrm{std}})$. This is useful for the following reason: $\mathbb{C}^2$ contains two obvious families of holomorphic planes

$$f_w : \mathbb{C} \to \mathbb{C}^2 : z \mapsto (z, w), \qquad \text{for } w \in \mathbb{C},$$
$$g_w : \mathbb{C} \to \mathbb{C}^2 : z \mapsto (w, z), \qquad \text{for } w \in \mathbb{C},$$

all of which are properly embedded maps, with two distinct types of asymptotic behavior. To describe the latter, choose a large constant $R > 0$, let $\mathbb{D}_R^4 \subset \mathbb{C}^2$ denote the disk of radius R and identify $\mathbb{C}^2 \backslash \mathbb{D}_R^4$ with $(R, \infty) \times S^3$ by viewing S^3 as the unit sphere in $\mathbb{C}^2$ and applying the diffeomorphism

$$(R, \infty) \times S^3 \xrightarrow{\cong} \mathbb{C}^2 \backslash \mathbb{D}_R^4 : (r, x) \mapsto rx.$$

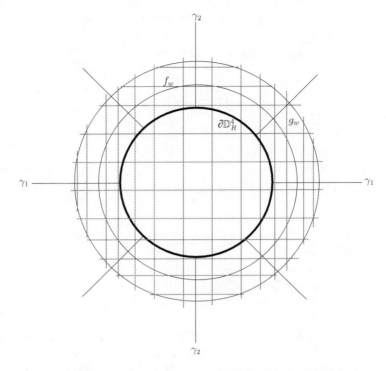

Figure 2 The two families of properly embedded holomorphic planes f_w and g_w form a coordinate grid for $\mathbb{C}^2$ and are each asymptotic on the cylindrical end $\mathbb{C}^2 \setminus \mathbb{D}_R^4 \cong (R, \infty) \times S^3$ to one of two specific loops $\gamma_1, \gamma_2 \subset S^3$.

Then each f_w or g_w maps a neighborhood of infinity into an arbitrarily small neighborhood of the cylinder $(R, \infty) \times \gamma_1$ or $(R, \infty) \times \gamma_2$, respectively, where we define

$$\gamma_1 := S^1 \times \{0\} \subset S^3 \subset \mathbb{C}^2, \qquad \gamma_2 := \{0\} \times S^1 \subset S^3 \subset \mathbb{C}^2.$$

A schematic picture of this asymptotic behavior and the resulting transverse pair of holomorphic foliations of $\mathbb{C}^2$ is shown in Figure 2. Informally, we will say that the planes f_w are asymptotic to γ_1 and the planes g_w are asymptotic to γ_2; more precise definitions of this terminology will appear in §2.4 when we discuss asymptotically cylindrical maps.

Now since $K' \subset \mathbb{C}^2 = \mathbb{R}^4$ is compact, $\mathbb{D}_R^4$ will contain K' for any $R > 0$ sufficiently large, so that we can also regard (M, ω) as containing a copy of the region identified above with $(R, \infty) \times S^3$. Let us fix such a radius and choose a compatible almost complex structure J on (M, ω) that matches the

Motivation

standard multiplication by i on $\mathbb{C}^2 \setminus \mathbb{D}_R^4 \cong (R, \infty) \times S^3$. The curves f_w and g_w can then be regarded as J-holomorphic planes in M for every $w \in \mathbb{C}$ with $|w| > R$, and just as in Theorem 1, these two families define elements in a pair of connected moduli spaces $\mathcal{M}_1(J; \gamma_1)$ and $\mathcal{M}_2(J; \gamma_2)$ of J-holomorphic planes in M, where we can use the loops γ_1 and γ_2 to prescribe the asymptotic behavior of the curves in the moduli spaces. There exists a well-developed theory of moduli spaces of J-holomorphic curves with this type of asymptotic behavior, a survey of which is given in Appendix A.2. In the present context, it can be applied to prove that $\mathcal{M}_1(J; \gamma_1)$ and $\mathcal{M}_2(J; \gamma_2)$ are both smooth 2-dimensional manifolds, and they are also compact except for the obvious way in which they are not: a sequence $u_j \in \mathcal{M}_k(J; \gamma_k)$ for $k \in \{1, 2\}$ will fail to have a convergent subsequence if and only if for large j it is of the form $u_j = f_{w_j} \in \mathcal{M}_1(J; \gamma_1)$ or $u_j = g_{w_j} \in \mathcal{M}_2(J; \gamma_2)$ for a sequence $w_j \in \mathbb{C}$ with $|w_j| \to \infty$. This gives each of $\mathcal{M}_1(J; \gamma_1)$ and $\mathcal{M}_2(J; \gamma_2)$ the topology of a compact surface with one boundary component attached to a *cylindrical end* of the form $\mathbb{C} \setminus \mathbb{D}_R \cong (R, \infty) \times S^1$.

If we want to apply these two moduli spaces the same way they were used in Theorem 1, then we need to establish the following:

Lemma 4 *The moduli spaces $\mathcal{M}_1(J; \gamma_1)$ and $\mathcal{M}_2(J; \gamma_2)$ described above have the following properties:*

(1) For each $k = 1, 2$ and every pair of distinct elements $u, v \in \mathcal{M}_k(J; \gamma_k)$, the images of $u \colon \mathbb{C} \to M$ and $v \colon \mathbb{C} \to M$ are disjoint.

(2) For every $u \in \mathcal{M}_1(J; \gamma_1)$ and $v \in \mathcal{M}_2(J; \gamma_2)$, the maps $u \colon \mathbb{C} \to M$ and $v \colon \mathbb{C} \to M$ have exactly one intersection point, which is transverse and positive.

(3) Every element of $\mathcal{M}_1(J; \gamma_1)$ or $\mathcal{M}_2(J; \gamma_2)$ is embedded.

Indeed, one can then argue exactly as in the proof of Theorem 1 that the two moduli spaces $\mathcal{M}_1(J; \gamma_1)$ and $\mathcal{M}_2(J; \gamma_2)$ form two transverse smooth foliations of M by planes, producing a coordinate grid (see Figure 3) that identifies M with $\mathbb{C} \times \mathbb{C} \cong \mathbb{R}^4$. The question I would now like to focus on is this: why is Lemma 4 true?

The answer does not come from homological intersection theory, as the curves in $\mathcal{M}_1(J; \gamma_1)$ and $\mathcal{M}_2(J; \gamma_2)$ are noncompact and do not represent homology classes. One can, however, use differential topological arguments to verify the second claim in the lemma: the fact that each f_w intersects each $g_{w'}$ exactly once transversely implies via a homotopy argument that the same will be true for any pair $u \in \mathcal{M}_1(J; \gamma_1)$ and $v \in \mathcal{M}_2(J; \gamma_2)$. Indeed, $\mathcal{M}_1(J; \gamma_1)$

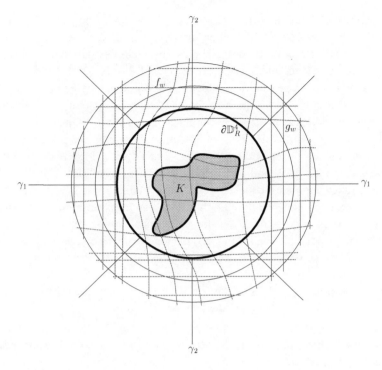

Figure 3 The moduli spaces $\mathcal{M}_1(J; \gamma_1)$ and $\mathcal{M}_2(J; \gamma_2)$ of proper J-holomorphic planes asymptotic to the loops $\gamma_1, \gamma_2 \subset S^3$ form two transverse foliations of M in Theorem 2, building a coordinate grid that proves $M \cong \mathbb{C} \times \mathbb{C} = \mathbb{R}^4$.

and $\mathcal{M}_2(J; \gamma_2)$ are each connected spaces of properly embedded planes that are asymptotic to disjoint loops in S^3, and thus they map neighborhoods of infinity to completely disjoint regions near infinity in M. This ensures that there exist homotopies of properly embedded maps

$$u_\tau : \mathbb{C} \to M, \qquad v_\tau : \mathbb{C} \to M, \qquad \tau \in [0, 1]$$

with $u_0 = u$, $u_1 = f_w$, $v_0 = v$ and $v_1 = g_{w'}$ such that the intersections of u_τ with v_τ for every $\tau \in [0, 1]$ are confined to compact subsets of both domains. Standard arguments as in [Mil97] then imply that u and v must have the same algebraic intersection count as f_w and $g_{w'}$, which is 1, so in light of positivity of intersections, u and v can only have one intersection point, and it must be transverse.

This type of argument does not suffice to prove the other two claims in Lemma 4. For example, suppose we would like to prove that two distinct curves $u, v \in \mathcal{M}_1(J; \gamma_1)$ must always be disjoint. It is easy to believe this in light of

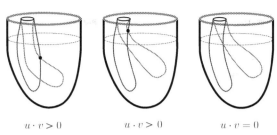

$u \cdot v > 0$ $u \cdot v > 0$ $u \cdot v = 0$

Figure 4 The algebraic intersection count $u \cdot v \in \mathbb{Z}$ between two proper maps of noncompact domains can change under homotopies if the two maps have matching asymptotic behavior.

the curves that we can explicitly see; i.e., f_w and $f_{w'}$ both belong to $\mathcal{M}_1(J; \gamma_1)$ for any $w, w' \in \mathbb{C}$ sufficiently large, and they are clearly disjoint if $w \neq w'$. To extend this to the curves that we cannot explicitly see because they do not live entirely in the region $(R, \infty) \times S^3 \subset M$, we would ideally like to argue via homotopy invariance, namely that if u_τ and v_τ are two continuous families of curves in $\mathcal{M}_1(J; \gamma_1)$ with u_0 and v_0 disjoint, then u_1 and v_1 must also be disjoint. But here we have a problem that did not arise in the previous paragraph: the curves u_τ and v_τ in this homotopy are always asymptotic to *the same* loop $\gamma_1 \subset S^3$, so their images in M always become arbitrarily close to each other in the cylindrical end $(R, \infty) \times S^3$. In this situation, there is no way to make sure that intersections are confined to compact subsets, and we can imagine, in fact, that under a homotopy, some intersections might just escape to infinity and disappear (see Figure 4)!

It is a remarkable fact that, in the situation under consideration, this nightmare scenario cannot happen, and Lemma 4 is indeed true. To understand why, we will have to explore the asymptotic behavior of noncompact J-holomorphic curves much more deeply. Still more interesting, perhaps, is that in more general situations, the nightmare scenario of Figure 4 really can happen, but it can also be *controlled*: one can define an *asymptotic contribution* that measures the possibility for "hidden" intersections to emerge from infinity under small perturbations. It turns out that just like the contribution of an isolated intersection between two J-holomorphic curves, this asymptotic contribution is always nonnegative, and adding it to the algebraic count of actual intersections produces a meaningful homotopy-invariant intersection product. Once this product and the corresponding generalization of the adjunction formula have been understood, proving results like Lemma 4 becomes quite easy.

The first hints of a systematic intersection theory for noncompact holomorphic curves appeared in Hutchings's work on embedded contact homology

10 *Motivation*

[Hut02], and the theory was developed in earnest a few years later in the Ph.D. thesis of Richard Siefring [Sie05] and his two papers [Sie08, Sie11]. Our primary objectives in these lectures will be to explain where this theory comes from, demonstrate how to use it, and give some examples of what it can be used for. We'll start in Lectures 1 and 2 by reviewing the intersection theory for closed holomorphic curves and discussing one of its most famous applications, McDuff's theorem [McD90] on symplectic ruled surfaces (which is a variation on Theorem 1). The asymptotic analysis required for Siefring's theory is then surveyed in Lecture 3 (mostly without the proofs since these are analytically somewhat intense), and Lecture 4 uses these asymptotic results to define the precise generalizations of the homological intersection product and the adjunction formula that are needed for results such as Lemma 4. In Lecture 5, we will demonstrate how to use the theory via a generalization of Theorem 2, framed in the language of contact 3-manifolds and their symplectic fillings.

Lecture 1

Closed Holomorphic Curves in Symplectic 4-Manifolds

In these lectures we would like to explain some results about symplectic 4-manifolds with contact boundary, and some of the technical tools involved in proving them, notably the intersection theory of punctured pseudoholomorphic curves. These tools are relatively recent but have historical precedents that go back to the late 1980s, when the field of symplectic topology was relatively new and many deep results about closed symplectic 4-manifolds were proved. We begin with a discussion of some of those results.

1.1 Some Examples of Symplectic 4-Manifolds

Suppose M is a smooth manifold of even dimension $2n \geq 2$. A **symplectic form** on M is a closed 2-form ω that is **nondegenerate**, meaning that $\omega(X, \cdot) \neq 0$ for every nonzero vector $X \in TM$, or equivalently,

$$\omega^n := \omega \wedge \cdots \wedge \neq 0$$

everywhere on M. This means that ω^n is a volume form, and thus it induces a natural orientation on M. We will always assume that any symplectic manifold (M, ω) carries the natural orientation induced by its symplectic structure, and thus we can write

$$\omega^n > 0.$$

We say that a submanifold $\Sigma \subset M$ is a **symplectic submanifold**, or is **symplectically embedded**, if $\omega|_{T\Sigma}$ is also nondegenerate.

Exercise 1.1 Show that every finite-dimensional manifold admitting a nondegenerate 2-form has even dimension.

12   *Closed Holomorphic Curves in Symplectic 4-Manifolds*

There are many interesting questions one can study on a symplectic manifold (M, ω); e.g., one can investigate the *Hamiltonian dynamics* for a function $H \colon M \to \mathbb{R}$, or one can study *symplectic embedding obstructions* of one symplectic manifold into another (see, e.g., [HZ94, MS17] for more on each of these topics). In this lecture, we will consider the most basic question of symplectic topology: given two closed symplectic manifolds (M, ω) and (M', ω') of the same dimension, what properties can permit us to conclude that they are **symplectomorphic**, i.e., there exists a diffeomorphism

$$\varphi \colon M \xrightarrow{\cong} M' \quad \text{with} \quad \varphi^* \omega' = \omega?$$

We shall deal with two fundamental examples of symplectic manifolds in dimension 4, of which the second is a generalization of the first.

Example 1.2   Suppose Σ is a closed, connected and oriented surface, and $\pi \colon M \to \Sigma$ is a smooth fiber bundle whose fibers are also closed, connected and oriented surfaces. The following result of Thurston says that under a mild (and obviously necessary) homological assumption, such fibrations always carry a canonical deformation class of symplectic forms.

Theorem 1.3 (Thurston [Thu76])   *Given a fibration $\pi \colon M \to \Sigma$ as described above, suppose the homology class of the fiber is not torsion in $H_2(M)$. Then M admits a symplectic form ω such that all fibers are symplectic submanifolds of (M, ω). Moreover, the space of symplectic forms on M having this property is connected.*

A symplectic manifold (M, ω) with a fibration whose fibers are symplectic is called a **symplectic fibration**. As a special case, if the fibers of $\pi \colon M \to \Sigma$ are spheres and Σ is a closed oriented surface, then a symplectic fibration (M, ω) over Σ is called a **symplectic ruled surface**. This term is inspired by complex algebraic geometry; in particular, the word "surface" refers to the fact that such manifolds can also be shown to admit complex structures, which makes them 2-dimensional complex manifolds, i.e., *complex surfaces*.

Exercise 1.4   Show that the homological condition in Theorem 1.3 is always satisfied if the fibers are spheres. *Hint:* $A \in H_2(M)$ is a torsion class if and only if the homological intersection number $A \cdot B \in \mathbb{Z}$ vanishes for all $B \in H_2(M)$. Consider the vertical subbundle $VM \subset TM \to M$, defined as the set of all vectors in TM that are tangent to fibers of $\pi \colon M \to \Sigma$. How many times (algebraically) does the zero-set of a generic section of $VM \to M$ intersect a generic fiber of $\pi \colon M \to \Sigma$?

The above class of examples is a special case of the following more general class.

1.1 Some Examples of Symplectic 4-Manifolds

Example 1.5 Suppose M and Σ are closed, connected, oriented, smooth manifolds of dimensions 4 and 2, respectively. A **Lefschetz fibration** of M over Σ is a smooth map

$$\pi : M \to \Sigma$$

with finitely many critical points $M^{\mathrm{crit}} := \mathrm{Crit}(\pi) \subset M$ and critical values $\Sigma^{\mathrm{crit}} := \pi(M^{\mathrm{crit}}) \subset \Sigma$ such that, near each point $p \in M^{\mathrm{crit}}$, there exists a complex coordinate chart (z_1, z_2) compatible with the orientation of M, and a corresponding complex coordinate z on a neighborhood of $\pi(p) \in \Sigma^{\mathrm{crit}}$ compatible with the orientation of Σ, in which π locally takes the form

$$\pi(z_1, z_2) = z_1^2 + z_2^2. \tag{1.1}$$

Remark 1.6 Any $2n$-dimensional manifold M admits a set of complex coordinates $(z_1, \ldots, z_n)$ near any point $p \in M$, but it is not always possible to cover M with such coordinate charts so that the transition maps are holomorphic; this is possible if and only if M also admits a **complex structure**. In the definition above, we have not assumed that M admits a complex structure, as the coordinates (z_1, z_2) are only required to exist locally near the finite set M^{crit}. Note, however, that any choice of complex coordinates on some domain determines an orientation on that domain: this follows from the fact that under the natural identification $\mathbb{R}^{2n} = \mathbb{C}^n$, any complex linear isomorphism $\mathbb{C}^n \to \mathbb{C}^n$, when viewed as an element of $\mathrm{GL}(2n, \mathbb{R})$, has positive determinant. In the above definition, we are assuming that the given orientations of M and Σ always match the orientations determined by the complex local coordinates.

A Lefschetz fibration restricts to a smooth fiber bundle over the set $\Sigma \backslash \Sigma^{\mathrm{crit}}$, and the fibers of this bundle are called the **regular fibers** of M; they are in general closed oriented surfaces, and we may always assume, without loss of generality, that they are connected (see Exercise 1.9 later). The finitely many **singular fibers** $\pi^{-1}(z)$ for $z \in \Sigma^{\mathrm{crit}}$ are immersed surfaces with finitely many double points that look like the transverse intersection of $\mathbb{C} \times \{0\}$ and $\{0\} \times \mathbb{C}$ in $\mathbb{C}^2$; one can see this by rewriting (1.1) in the coordinates $\zeta_1 := z_1 + iz_2$ and $\zeta_2 := z_1 - iz_2$, so that the local model becomes $\pi(\zeta_1, \zeta_2) = \zeta_1 \zeta_2$. Each singular fiber is uniquely decomposable into a transversely intersecting union of subsets that are immersed images of *connected* surfaces: we call these subsets the **irreducible components** (see Figure 1.1).

Thurston's theorem about symplectic structures on fibrations was generalized to Lefschetz fibrations by Gompf. To state the most useful version of this result, we need to generalize the notion of a "symplectic submanifold" in a way that will also make sense for singular fibers, which are not embedded

14   *Closed Holomorphic Curves in Symplectic 4-Manifolds*

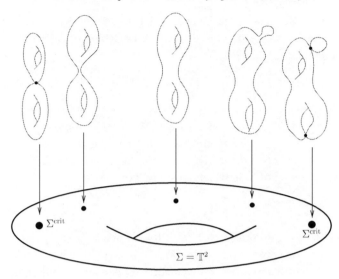

Figure 1.1 A Lefschetz fibration over $\mathbb{T}^2$ with regular fibers of genus 2 and two singular fibers, each of which has two irreducible components.

submanifolds. Since Lefschetz critical points are defined in terms of complex local coordinates, one way to do this is by elucidating the relationship between complex and symplectic structures.

Definition 1.7 Suppose $E \to B$ is a smooth real vector bundle of even rank. A **complex structure** on $E \to B$ is a smooth linear bundle map $J: E \to E$ such that $J^2 = -\mathbb{1}$. A **symplectic structure** on $E \to B$ is a smooth antisymmetric bilinear bundle map $\omega: E \oplus E \to \mathbb{R}$ that is nondegenerate, meaning $\omega(v, \cdot) \neq 0$ for all nonzero $v \in E$. We say that ω **tames** J if for all $v \in E$ with $v \neq 0$, we have
$$\omega(v, Jv) > 0.$$
We say additionally that J is **compatible** with ω if the pairing
$$g_J(v, w) := \omega(v, Jw)$$
is both nondegenerate and symmetric; i.e., it defines a bundle metric.

One can show that a complex or symplectic structure on a vector bundle implies the existence of local trivializations for which all transition maps are complex linear maps $\mathbb{C}^n \to \mathbb{C}^n$ or symplectic linear maps $\mathbb{R}^{2n} \to \mathbb{R}^{2n}$, respectively (see [MS17] for details). An **almost complex structure** on a manifold M is simply a complex structure on its tangent bundle $TM \to M$. Here the word

1.1 Some Examples of Symplectic 4-Manifolds

"almost" is inserted in order to distinguish this relatively weak notion from the much more rigid notion mentioned in Remark 1.6: a complex manifold carries a natural almost complex structure (defined via multiplication by i in any holomorphic coordinate chart), but not every almost complex structure arises in this way from local charts, and there are many manifolds that admit *almost* complex structures but not complex structures. One way to paraphrase Definition 1.7 is to say that ω tames J if and only if every complex 1-dimensional subspace of a fiber in E is also a *symplectic subspace*; similarly, if (M, ω) is a symplectic manifold, then ω tames an almost complex structure J on M if and only if every *complex curve* in the *almost complex manifold* (M, J) is also a symplectic submanifold.

With this understood, suppose $\pi: M \to \Sigma$ is a Lefschetz fibration as defined above. We will say that a symplectic form ω on M is **supported by** π if the following conditions hold:

(1) Every fiber of $\pi|_{M \setminus M^{\mathrm{crit}}}: M \setminus M^{\mathrm{crit}} \to \Sigma$ is a symplectic submanifold.
(2) On a neighborhood of M^{crit}, ω tames some almost complex structure J that preserves the tangent spaces of the fibers.

Gompf's generalization of Thurston's theorem can now be stated as follows:

Theorem 1.8 (Gompf [GS99]) *Suppose M and Σ are closed, connected and oriented manifolds of dimensions 4 and 2, respectively, and $\pi: M \to \Sigma$ is a Lefschetz fibration for which the fiber represents a nontorsion class in $H_2(M)$. Then the space of symplectic forms on M that are supported by π is nonempty and connected.*

A Lefschetz fibration $\pi: M \to \Sigma$ on a symplectic manifold (M, ω) with ω supported in the above sense is called a **symplectic Lefschetz fibration**.

Exercise 1.9 Assuming M and Σ are closed and connected, show that if $\pi: M \to \Sigma$ is a Lefschetz fibration with disconnected fibers, then one can write $\pi = \varphi \circ \pi'$ where $\varphi: \Sigma' \to \Sigma$ is a finite covering map of degree at least 2 and $\pi': M \to \Sigma'$ is a Lefschetz fibration with connected fibers.

There is a natural way to replace any smooth fiber bundle as in Example 1.2 with a Lefschetz fibration that has singular fibers, namely by *blowing up* finitely many points. Topologically, this can be described as follows: given $p \in M$, choose local complex coordinates (z_1, z_2) on some neighborhood $\mathcal{N}(p) \subset M$ of p that are compatible with the orientation and identify p with $0 \in \mathbb{C}^2$. Let $E \to \mathbb{CP}^1$ denote the tautological complex line bundle; i.e., the bundle whose fiber over $[z_1 : z_2] \in \mathbb{CP}^2$ is the complex line spanned by $(z_1, z_2) \in \mathbb{C}^2$. There is a canonical identification of $E \setminus \mathbb{CP}^1$ with $\mathbb{C}^2 \setminus \{0\}$, where $\mathbb{CP}^1 \subset E$ here denotes

16   *Closed Holomorphic Curves in Symplectic 4-Manifolds*

the zero-section. Thus for some neighborhood $\mathcal{N}(\mathbb{CP}^1) \subset E$ of $\mathbb{CP}^1$, the above coordinates allow us to identify $\mathcal{N}(p)\backslash\{p\}$ with $\mathcal{N}(\mathbb{CP}^1)\backslash\mathbb{CP}^1$, and we define the (smooth, oriented) **blowup** $\widehat{M}$ of M by removing $\mathcal{N}(p)$ and replacing it with $\mathcal{N}(\mathbb{CP}^1)$. There is a natural projection

$$\Phi \colon \widehat{M} \to M$$

such that $S := \Phi^{-1}(p)$ is a smoothly embedded 2-sphere $S \cong \mathbb{CP}^1 \subset \widehat{M}$ (called an **exceptional sphere**), whose homological self-intersection number satisfies

$$[S] \cdot [S] = -1. \tag{1.2}$$

The restriction of Φ to $\widehat{M}\backslash S$ is a diffeomorphism onto $M\backslash\{p\}$.

Exercise 1.10 Show that if $\pi \colon M \to \Sigma$ is a Lefschetz fibration and $p \in M\backslash M^{\mathrm{crit}}$, then there exist complex local coordinates (z_1, z_2) for a neighborhood of p in M and z for a neighborhood of $\pi(p)$ in Σ, both compatible with the orientations, such that π takes the form $\pi(z_1, z_2) = z_1$ near p.

Exercise 1.11 Suppose $\pi \colon M \to \Sigma$ is a Lefschetz fibration, and $\widehat{M}$ is obtained by blowing up M at a point $p \in M\backslash M^{\mathrm{crit}}$, using a complex coordinate chart as in Exercise 1.10. Then if $\Phi \colon \widehat{M} \to M$ denotes the induced projection map, show that $\pi \circ \Phi \colon \widehat{M} \to \Sigma$ is a Lefschetz fibration, having one more critical point than $\pi \colon M \to \Sigma$ and containing the exceptional sphere $\Phi^{-1}(p)$ as an irreducible component of a singular fiber.

Exercise 1.12 Prove that the sphere $S \subset \widehat{M}$ created by blowing up M at a point satisfies (1.2). *Hint:* You only need to know the first Chern number of the tautological line bundle.

Exercise 1.13 Prove that if $\widehat{M}$ is constructed by blowing up M at a point, then $\widehat{M}$ is diffeomorphic to the connected sum $M\#\overline{\mathbb{CP}^2}$, where the line over $\mathbb{CP}^2$ indicates that it carries the opposite of its canonical orientation (determined by the complex structure of $\mathbb{CP}^2$). *Hint:* Present $\mathbb{CP}^2$ as the union of $\mathbb{C}^2$ with a "sphere at infinity" $\mathbb{CP}^1 \subset \mathbb{CP}^2$. What does a tubular neighborhood of $\mathbb{CP}^1$ in $\mathbb{CP}^2$ look like, and what changes if you reverse the orientation?

It is easy to prove, from the above description of the blowup, that if M is a complex manifold, $\widehat{M}$ inherits a canonical complex structure. What is somewhat less obvious, but nonetheless true and hopefully not so surprising by this point, is that if (M, ω) is symplectic, then $\widehat{M}$ also inherits a symplectic form $\widehat{\omega}$ that is canonical up to smooth deformation through symplectic forms (see [MS17] or [Wen18, §3.2]). In this case, the resulting exceptional sphere is a symplectic submanifold of $(\widehat{M}, \widehat{\omega})$. Conversely, if (M, ω) is any symplectic 4-manifold containing a symplectically embedded exceptional sphere $S \subset M$,

1.2 McDuff's Characterization of Symplectic Ruled Surfaces 17

then one can reverse the above operation and show that (M, ω) is the symplectic blowup of another symplectic manifold (M_0, ω_0), with the resulting projection $\Phi \colon M \to M_0$ collapsing S to a point. We say that a symplectic 4-manifold is **minimal** if it contains no symplectically embedded exceptional spheres, which means it is not the blowup of any other symplectic manifold. McDuff [McD90] proved the following theorem:

Theorem 1.14 (McDuff [McD90]) *If (M, ω) is a closed symplectic 4-manifold with a maximal collection of pairwise disjoint exceptional spheres $E_1, \ldots, E_N \subset (M, \omega)$, then the symplectic manifold obtained from (M, ω) by "blowing down" along $E_1, \ldots, E_N$ is minimal.*

One can also show that if ω is supported by a Lefschetz fibration $\pi \colon M \to \Sigma$, then the symplectic form $\widehat{\omega}$ on the blowup $\widehat{M}$ can be arranged to be supported by the Lefschetz fibration on $\widehat{M}$ arising from Exercise 1.11 (see, e.g., [Wen18, Theorem 3.44]).

Symplectic fibrations are a rather special class of symplectic 4-manifolds, but the following deep theorem of Donaldson indicates that Lefschetz fibrations are surprisingly general examples. The theorem is actually true in all dimensions; we will not make use of it in any concrete way in these notes, but it is important to have as a piece of background knowledge.

Theorem 1.15 (Donaldson [Don99]) *Any closed symplectic manifold can be blown up finitely many times to a symplectic manifold that admits a symplectic Lefschetz fibration over S^2.*

1.2 McDuff's Characterization of Symplectic Ruled Surfaces

If (M, ω) is a symplectic 4-manifold with a supporting Lefschetz fibration $\pi \colon M \to \Sigma$, then it admits a 2-dimensional symplectic submanifold $S \subset (M, \omega)$ satisfying

$$[S] \cdot [S] = 0;$$

indeed, S can be chosen to be any regular fiber of the Lefschetz fibration. The following remarkable result says that if S has genus 0, then the converse also holds.

Theorem 1.16 (McDuff [McD90]) *Suppose (M, ω) is a closed and connected symplectic 4-manifold, and $S \subset M$ is a symplectically embedded 2-sphere satisfying $[S] \cdot [S] = 0$. Then S is a fiber of a symplectic Lefschetz fibration $\pi \colon M \to \Sigma$ over some closed oriented surface Σ, and π is a smooth symplectic*

18 *Closed Holomorphic Curves in Symplectic 4-Manifolds*

fibration (i.e., without Lefschetz critical points) whenever $(M \setminus S, \omega)$ is minimal. In particular, (M, ω) can be obtained by blowing up a symplectic ruled surface finitely many times.

This theorem is false for surfaces S with positive genus (see Remark A.9 for more on this). There is also no comparably strong result about symplectic fibrations in dimensions greater than 4, as the theory of holomorphic curves is considerably stronger in low dimensions. Our main goal for the rest of this lecture will be to sketch a proof of the theorem.

The proof begins with the observation, originally due to Gromov [Gro85], that every symplectic manifold (M, ω) admits an almost complex structure J that is *compatible* with ω in the sense of Definition 1.7. Moreover, if $S \subset (M, \omega)$ is a symplectic submanifold, one can easily choose a compatible almost complex structure J that preserves TS; i.e., it makes S into an embedded J-complex curve. The main idea of the proof is then to study the entire space of J-complex curves homologous to S and show that these must foliate M, possibly with finitely many singularities.

Let us define the "space of J-complex curves" more precisely. Recall that a **Riemann surface** can be regarded as an almost complex[1] manifold (Σ, j) with[2] $\dim \Sigma = 2$. Given (Σ, j) and an almost complex manifold (M, J) of real dimension $2n$, we say that a smooth map $u \colon \Sigma \to M$ is J-**holomorphic**, or **pseudoholomorphic** (often abbreviated simply as "holomorphic"), if its tangent map is complex linear at every point; i.e.,

$$Tu \circ j \equiv J \circ Tu. \tag{1.3}$$

This is a first-order elliptic PDE: in any choice of holomorphic local coordinates $s + it$ on a domain in Σ, (1.3) is equivalent to the nonlinear Cauchy–Riemann type equation

$$\partial_s u(s, t) + J(u(s, t)) \, \partial_t u(s, t) = 0.$$

Solutions are called **pseudoholomorphic curves**, where the word "curve" refers to the fact that their domains are complex *1*-dimensional manifolds. They have many nice properties, which are proved by a combination of complex function theory, nonlinear functional analysis and elliptic regularity theory – a quick overview of the essential properties is given in Appendix A, and some of these will be used in the following discussion.

---

[1] Due to a theorem of Gauss, every almost complex structure on a manifold of real dimension 2 is *integrable*; i.e., it arises from an atlas of coordinate charts with holomorphic transition maps and is thus also a complex structure (without the "almost").

[2] Unless otherwise noted, all dimensions mentioned in this book will be *real* dimensions, not complex.

1.2 McDuff's Characterization of Symplectic Ruled Surfaces   19

For any integer $g \geq 0$ and $A \in H_2(M)$, we define the **moduli space** $\mathcal{M}_g^A(M,J)$ of **unparametrized closed J-holomorphic curves** of genus g homologous to A as the space of equivalence classes $[(\Sigma, j, u)]$, where (Σ, j) is a closed connected Riemann surface of genus g, $u\colon (\Sigma, j) \to (M, J)$ is a pseudoholomorphic map representing the homology class $[u] := u_*[\Sigma] = A$, and we write $(\Sigma, j, u) \sim (\Sigma', j', u')$ if and only if they are related to each other by reparametrization; i.e., there exists a holomorphic diffeomorphism $\varphi\colon (\Sigma, j) \to (\Sigma', j')$ (a **biholomorphic** map) such that $u = u' \circ \varphi$. We will sometimes abuse notation and abbreviate an equivalence class $[(\Sigma, j, u)] \in \mathcal{M}_g^A(M, J)$ simply as the parametrization "u" when there is no danger of confusion. The notion of C^∞-convergence defines a natural topology on $\mathcal{M}_g^A(M, J)$ such that a sequence $[(\Sigma_k, j_k, u_k)] \in \mathcal{M}_g^A(M, J)$ converges to $[(\Sigma, j, u)] \in \mathcal{M}_g^A(M, J)$ if and only if there exist representatives $(\Sigma, j_k', u_k') \sim (\Sigma_k, j_k, u_k)$ for which

$$j_k' \to j \quad \text{and} \quad u_k' \to u$$

uniformly with all derivatives on Σ. In cases where we'd prefer not to specify the homology class, we will occasionally write

$$\mathcal{M}_g(M, J) := \coprod_{A \in H_2(M)} \mathcal{M}_g^A(M, J).$$

Observe that if $u\colon (\Sigma, j) \to (M, J)$ is a closed J-holomorphic curve and $\varphi\colon (\Sigma', j') \to (\Sigma, j)$ is a holomorphic map from another closed Riemann surface (Σ', j'), then $u \circ \varphi\colon (\Sigma', j') \to (M, J)$ is also a J-holomorphic curve. If φ is nonconstant, then holomorphicity implies that it has degree $\deg(\varphi) \geq 1$, with equality if and only if it is biholomorphic; in the case $k := \deg(\varphi) > 1$, we then say that u' is a k**-fold multiple cover** of u. Note that in this situation $[u'] = k[u]$, for instance, a curve cannot be a multiple cover if it represents a primitive homology class. We say that a nonconstant closed J-holomorphic curve is **simple** if it is not a multiple cover of any other curve.

Returning to the specific situation of McDuff's theorem, assume J is an ω-compatible almost complex structure that preserves the tangent spaces of the symplectically embedded sphere $S \subset (M, \omega)$. Then $(S, J|_{TS})$ is a closed Riemann surface of genus 0, and its inclusion $u_S\colon S \hookrightarrow M$ is an embedded J-holomorphic curve, defining an element

$$u_S \in \mathcal{M}_0^{[S]}(M, J)$$

in the moduli space of J-holomorphic spheres homologous to S. A straightforward application of standard machinery now gives the following result, a proof of which may be found at the end of Appendix A.1.

20   *Closed Holomorphic Curves in Symplectic 4-Manifolds*

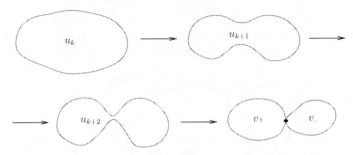

Figure 1.2 A sequence of J-holomorphic spheres u_k degenerating to a nodal curve $\{v_+, v_-\}$.

Lemma 1.17 *After a C^∞-small perturbation of J outside a neighborhood of S, the open subset $\mathcal{M}_0^{[S],*}(M, J) \subset \mathcal{M}_0^{[S]}(M, J)$, consisting of simple J-holomorphic spheres homologous to $[S]$, is a smooth oriented 2-dimensional manifold, and it is "compact up to bubbling" in the following sense. There exists a finite set of simple curves $\mathcal{B} \subset \mathcal{M}_0(M, J)$ with positive first Chern numbers such that if $u_k \in \mathcal{M}_0^{[S],*}(M, J)$ is a sequence with no convergent subsequence in $\mathcal{M}_0^{[S]}(M, J)$, then it has a subsequence that degenerates (see Figure 1.2) to a nodal curve $\{v_+, v_-\} \in \overline{\mathcal{M}}_0^{[S]}(M, J)$ for some $v_+, v_- \in \mathcal{B}$.*

The above formulation is a bit lazy, since we have not as yet given any definition of the space $\overline{\mathcal{M}}_0(M, J)$ of *nodal curves*. More precise details of this *compactification* of $\mathcal{M}_0(M, J)$ may be found in Appendix A.1, but for the purposes of the present discussion, it will suffice to characterize the degeneration of a sequence $[(S^2, j_k, u_k)] \in \mathcal{M}_0^A(M, J)$ to a nodal curve $\{[(S^2, j_+, v_+)], [(S^2, j_-, v_-)]\} \in \overline{\mathcal{M}}_0^A(M, J)$ as follows. The nodal curve is assumed to have the property that any choice of representatives $(S^2, j_\pm, v_\pm)$ comes with a distinguished intersection

$$v_+(z_+) = v_-(z_-),$$

for some pair of points $z_\pm \in S^2$; this intersection is called the **node**. Given these parametrizations, let $C \subset S^2$ denote the equator of the sphere, separating it into the two hemispheres

$$S^2 = \mathcal{D}_+ \cup_C \mathcal{D}_-,$$

and choose continuous surjections $\varphi_\pm : \mathcal{D}_\pm \to S^2$ that map $\mathring{\mathcal{D}}_\pm$ diffeomorphically to $S^2 \setminus \{z_\pm\}$ and collapse C to $z_\pm$. The map

$$u_\infty : S^2 \to M : z \mapsto \begin{cases} v_- \circ \varphi_-(z) & \text{for } z \in \mathcal{D}_- \\ v_+ \circ \varphi_+(z) & \text{for } z \in \mathcal{D}_+ \end{cases}$$

1.2 McDuff's Characterization of Symplectic Ruled Surfaces   21

is then continuous and smooth on $S^2 \backslash C$. This also defines a complex structure on $S^2 \backslash C$ by

$$j_\infty := \begin{cases} \varphi_-^* j_- & \text{on } \mathring{D}_-, \\ \varphi_+^* j_+ & \text{on } \mathring{D}_+, \end{cases}$$

though j_∞ does not extend smoothly over C. Now the convergence $u_k \to \{v_+, v_-\}$ can be defined to mean that all of the above choices can be made together with choices of representatives (S^2, j_k, u_k) such that

$$u_k \to u_\infty \quad \text{in } C^0(S^2, M) \text{ and } C^\infty_{\text{loc}}(S^2 \backslash C, M),$$

and

$$j_k \to j_\infty \quad \text{in } C^\infty_{\text{loc}}(S^2 \backslash C).$$

Observe that as a result of the C^0-convergence, $[v_+] + [v_-] = A \in H_2(M)$.

Lemma 1.17 relies on very general properties of J-holomorphic curves that are valid in all dimensions; under a few extra assumptions, some version of the same result could be proved for a $2n$-dimensional symplectic manifold (M, ω) containing a symplectically embedded 2-sphere $S \subset M$ with trivial normal bundle. The following improvement, which we will prove in Lecture 2 (see §2.2), is unique to dimension 4:

Lemma 1.18 *The finitely many nodal curves*

$$\{v_+^1, v_-^1\}, \ldots, \{v_+^N, v_-^N\} \in \overline{\mathcal{M}}_0^{[S]}(M, J)$$

appearing in Lemma 1.17 have the following properties:

(1) *Each $v_\pm^i : S^2 \to M$ for $i = 1, \ldots, N$ is embedded and satisfies $[v_\pm^i] \cdot [v_\pm^i] = -1$.*
(2) *v_+^i and v_-^i for $i = 1, \ldots, N$ intersect each other exactly once, transversely.*
(3) *For $i, j \in \{1, \ldots, N\}$ with $i \neq j$,*

$$v_i^+(S^2) \cap v_j^+(S^2) = v_i^+(S^2) \cap v_j^-(S^2) = v_i^-(S^2) \cap v_j^-(S^2) = \emptyset.$$

Moreover, if $F \subset M$ denotes the union of all the images of these nodal curves, then the curves in $\mathcal{M}_0^{[S]}(M, J)$ are all embedded and have pairwise disjoint images that define a smooth foliation of some open subset of $M \backslash F$.

With this lemma at our disposal, the proof of Theorem 1.16 concludes as follows: let

$$X := \left\{ p \in M \backslash F \mid p \text{ is in the image of a curve in } \mathcal{M}_0^{[S]}(M, J) \right\}.$$

Lemma 1.18 guarantees that X is an open subset of $M \backslash F$, but by the compactness statement in Lemma 1.17, X is also a closed subset. Since $M \backslash F$

22   *Closed Holomorphic Curves in Symplectic 4-Manifolds*

is connected, we conclude that the curves in $\mathcal{M}_0^{[S]}(M, J)$ fill *all* of it. Now, the compactified moduli space $\overline{\mathcal{M}}_0^{[S]}(M, J)$ consists of $\mathcal{M}_0^{[S]}(M, J)$ plus finitely many additional elements in the form of nodal curves; it has the topology of some compact oriented 2-manifold Σ, and the above argument shows that every point in M is in the image of precisely one element of $\overline{\mathcal{M}}_0^{[S]}(M, J)$. This defines a map

$$\pi \colon M \to \overline{\mathcal{M}}_0^{[S]}(M, J) \cong \Sigma,$$

whose regular fibers are the images of the smoothly embedded curves in $\mathcal{M}_0^{[S]}(M, J)$, and the images of nodal curves give rise to Lefschetz singular fibers, each with a unique critical point where two embedded J-holomorphic spheres intersect transversely. Since all the fibers are images of J-holomorphic curves and J is ω-tame, the fibers are also symplectic submanifolds. Furthermore, the irreducible components of the singular fibers are exceptional spheres that are disjoint from S (since the latter is also a fiber); thus no singular fibers can exist if $(M \backslash S, \omega)$ is minimal.

Remark 1.19   One can also prove the converse of the statement about minimality; i.e., if the Lefschetz fibration has no singular fibers, then $(M \backslash S, \omega)$ must be minimal. This relies on another theorem of McDuff [McD90], that for generic J, any exceptional sphere is homologous to a unique J-holomorphic sphere, which is embedded. A more comprehensive exposition of this topic and the more general version of McDuff's theorem for *rational and ruled* symplectic 4-manifolds are given in [Wen18]; see also [LM96].

1.3 Local Foliations by Holomorphic Spheres

The distinctive power of holomorphic curve methods in dimension 4 results from the numerical coincidence that $2 + 2 = 4$: in particular, any pair of holomorphic curves $u \in \mathcal{M}_g^A(M, J)$ and $v \in \mathcal{M}_{g'}^{A'}(M, J)$ has a well-defined homological intersection number $[u] \cdot [v] = A \cdot A' \in \mathbb{Z}$. We will discuss this subject in earnest in §2.1, but before that, let us examine a slightly simpler phenomenon that is also distinctive to dimension 4 and important for the proof of Lemma 1.18.

Suppose (M, J) is a $2n$-dimensional almost complex manifold and $u \in \mathcal{M}_0^A(M, J)$ is an embedded J-holomorphic curve such that the normal bundle $N_u \to S^2$ to any parametrization $u \colon S^2 \hookrightarrow M$ is trivial. Since $du(z) \colon (T_z S^2, j) \to (T_{u(z)} M, J)$ is complex linear and injective for all $z \in S^2$, the normal bundle naturally inherits a complex structure such that

1.3 Local Foliations by Holomorphic Spheres

$$u^*TM \cong TS^2 \oplus N_u$$

as complex vector bundles, so the first Chern numbers of these bundles satisfy

$$c_1(u^*TM) = c_1(TS^2) + c_1(N_u) = \chi(S^2) + 0 = 2,$$

where $c_1(u^*TM)$ is shorthand for evaluation of $c_1(u^*TM, J) \in H^2(S^2)$ on the fundamental class:

$$c_1(u^*TM) := \langle c_1(u^*TM, J), [S^2] \rangle = \langle u^*c_1(TM, J), [S^2] \rangle$$
$$= \langle c_1(TM, J), u_*[S^2] \rangle =: c_1(A).$$

If $\dim M = 4$, then triviality of N_u implies that $u(S^2)$ is a symplectically embedded sphere with self-intersection number 0, and we saw in Lemma 1.17 that in this case $\dim \mathcal{M}_0^A(M, J) = 2$. More generally, plugging $\dim M = 2n$ and $c_1(A) = 2$ into the virtual dimension formula (A.1) in Appendix A.1 gives

$$\text{vir-dim}\, \mathcal{M}_0^A(M, J) = 2(n - 3) + 2c_1(A) = 2n - 2.$$

This means more precisely that if J is sufficiently generic, then the open subset of $\mathcal{M}_0^A(M, J)$ consisting only of simple curves is a smooth manifold of this dimension, and since u itself is embedded, this is true in particular for some neighborhood of u in $\mathcal{M}_0^A(M, J)$. Note also that embeddedness of spheres in M is an open condition, so all other curves near u are also embedded. This observation and the dimension computation above make the following question reasonable:

Question 1.20 *Do the curves near u in $\mathcal{M}_0^A(M, J)$ foliate a neighborhood of $u(S^2)$?*

To answer this, let us choose a Riemannian metric on M and assume there exists a smooth family of parametrizations for the curves near u via sections of its normal bundle; i.e., one can find a smooth map

$$\Psi : \mathbb{D}^{2n-2} \times S^2 \to M : (\sigma, z) \mapsto u_\sigma(z) := \exp_{u(z)} h_\sigma(z) \tag{1.4}$$

with $h_\sigma \in \Gamma(N_u)$ for each $\sigma \in \mathbb{D}^{2n-2}$ and $h_0 \equiv 0$ such that the maps u_σ parametrize curves in $\mathcal{M}_0^A(M, J)$ and $u_0 = u$. There is then a linear map $\mathbb{R}^{2n-2} \to \Gamma(N_u) : X \mapsto \eta_X$ defined by

$$\eta_X(z) = d\Psi(0, z)(X, 0) = \frac{d}{dt} u_{tX}(z) \Big|_{t=0},$$

and the image of this map can be identified with the tangent space $T_u \mathcal{M}_0^A(M, J)$. Using the fact that each $u_\sigma : S^2 \to M$ satisfies a nonlinear Cauchy–Riemann type equation, one can show that all the sections η_X satisfy some *linearized* Cauchy–Riemann type equation (cf. Appendix B.1.1): in particular,

24 *Closed Holomorphic Curves in Symplectic 4-Manifolds*

for any choice of local holomorphic coordinates $s + it$ identifying a domain $\mathcal{U} \subset \Sigma$ with some open set $\Omega \subset \mathbb{C}$, the local expression for η_X in a complex trivialization over $\mathcal{U}$ is a function $f \colon \Omega \to \mathbb{C}^{n-1}$ satisfying a linear PDE of the form

$$\partial_s f(s,t) + i \, \partial_t f(s,t) + A(s,t)f(s,t) = 0, \tag{1.5}$$

for some smooth function $A(s,t)$ valued in the space $\mathrm{End}_{\mathbb{R}}(\mathbb{C}^{n-1})$ of real-linear maps on $\mathbb{C}^{n-1}$. Except for the extra 0th-order term, this is the standard Cauchy–Riemann equation, and we might therefore expect f to have similar properties to an analytic function $\Omega \to \mathbb{C}^{n-1}$; e.g., its zeroes should be isolated unless $\eta_X \equiv 0$. This intuition is made precise by the following consequence of elliptic regularity theory, often called the **similarity principle**; a slight generalization of this result is stated and proved in Appendix B.1.

Theorem 1.21 (*similarity principle*) *Suppose $\Omega \subset \mathbb{C}$ is an open set, $N \in \mathbb{N}$, $A \colon \Omega \to \mathrm{End}_{\mathbb{R}}(\mathbb{C}^N)$ is smooth, $f \colon \Omega \to \mathbb{C}^N$ is a smooth function satisfying the equation (1.5), and $z_0 \in \Omega$ is a point with $f(z_0) = 0$. Then f can be written on some neighborhood $z_0 \in \mathcal{U} \subset \Omega$ as*

$$f(z) = \Phi(z)g(z), \qquad z \in \mathcal{U}, \tag{1.6}$$

for some continuous function $\Phi \colon \mathcal{U} \to \mathrm{End}_{\mathbb{C}}(\mathbb{C}^N)$ with $\Phi(z_0) = \mathbb{1}$ and a holomorphic function $g \colon \mathcal{U} \to \mathbb{C}^N$. Moreover, if A is complex linear at every point, then Φ can be taken to be smooth.

Corollary 1.22 *Given $f \colon \Omega \to \mathbb{C}^N$ as in Theorem 1.21, f is either identically zero or has only isolated zeroes. In the latter case, if $N = 1$, all zeroes of f have positive order.*

Proof Writing $f(z) = \Phi(z)g(z)$ as in (1.6) for z in a neighborhood $\mathcal{U}$ of z_0, we can assume after shrinking $\mathcal{U}$ that $\Phi(z)$ is close to $\mathbb{1}$ and thus invertible for all $z \in \mathcal{U}$. Then $f|_{\mathcal{U}}$ is identically zero if and only if $g|_{\mathcal{U}}$ is; otherwise, g has an isolated zero at z_0, and thus so does f. If the latter holds and also $N = 1$, then we can further conclude that the winding number of the loop

$$\mathbb{R}/\mathbb{Z} \to \mathbb{C}\backslash\{0\} \colon \theta \mapsto g(z_0 + \epsilon e^{2\pi i\theta})$$

for small $\epsilon > 0$ is positive, and since Φ is close to the identity, the same is true for f. $\qquad\square$

The similarity principle implies that sections $\eta_X \in T_u \mathcal{M}_0^A(M, J)$ have at most finitely many zeroes in general, but it implies much more than this in the case where $\dim M = 4$. Indeed, $N_u \to S^2$ is in this case a complex *line* bundle, so for any section of this bundle with only isolated zeroes, the algebraic count

1.3 Local Foliations by Holomorphic Spheres

of the zeroes is given by the first Chern number $c_1(N_u) \in \mathbb{Z}$, which vanishes since the bundle is trivial. But by Corollary 1.22, the zeroes of any nontrivial section $\eta_X \in T_u \mathcal{M}_0^A(M, J)$ all count positively, so it follows that there cannot be any: η_X is nowhere zero! This is true for all $X \neq 0$ and thus implies that $d\Psi(0, z): T_{(0,z)}(\mathbb{D}^2 \times S^2) \to T_{u(z)}M$ is an isomorphism for all $z \in S^2$; hence the map (1.4) is an embedding in some neighborhood of $\{0\} \times S^2$, giving a positive answer to Question 1.20.

Proposition 1.23 *If* $\dim M = 4$ *and* $u \in \mathcal{M}_0^A(M, J)$ *is an embedded J-holomorphic sphere with trivial normal bundle, then the images of the curves in* $\mathcal{M}_0^A(M, J)$ *near u foliate a neighborhood of the image of u.* $\square$

No such general result is possible when $\dim M > 4$, because there is no way to "count" the number of zeroes of a section of a higher-rank complex vector bundle over S^2.

Exercise 1.24 Suppose $L \to \Sigma$ is a complex line bundle over a closed Riemann surface (Σ, j), and $V \subset \Gamma(L)$ is a vector space of sections that satisfy a real-linear Cauchy–Riemann type equation, so in particular the similarity principle holds for sections $\eta \in V$. Prove $\dim_{\mathbb{R}} V \leq 2 + 2c_1(L)$.

Lecture 2

Intersections, Ruled Surfaces, and Contact Boundaries

In this lecture we explain the intersection theory for closed holomorphic curves in dimension 4 and use it to complete the overview from Lecture 1 of McDuff's theorem on ruled surfaces. We will then discuss the generalization of these ideas to punctured holomorphic curves in symplectic cobordisms and consider some applications to the study of symplectic fillings.

2.1 Positivity of Intersections and the Adjunction Formula

To complete the proof of Lemma 1.18 from §1.2, we must discuss the intersection theory of J-holomorphic curves in dimension 4. The notion of "homological" intersection numbers was mentioned already a few times in the previous lecture, and it will be useful now to review precisely what this means. Suppose M is a closed oriented smooth 4-manifold, Σ and Σ' are closed oriented surfaces, and

$$u \colon \Sigma \to M, \qquad v \colon \Sigma' \to M$$

are C^1-smooth maps. An intersection $u(z) = v(\zeta) = p$ is **transverse** if

$$\operatorname{im} \, du(z) \oplus \operatorname{im} \, dv(\zeta) = T_p M \tag{2.1}$$

and **positive** if and only if the natural orientation induced on this direct sum by the orientations of $T_z\Sigma$ and $T_\zeta\Sigma'$ matches the orientation of $T_p M$. Otherwise, it is called **negative**, and we define the **local intersection index** accordingly as $\iota(u, z\,;\, v, \zeta) = \pm 1$. If all intersections between u and v are transverse, then they are all isolated, and thus there are only finitely many, so we can define the total **intersection number**

$$[u] \cdot [v] := \sum_{u(z)=v(\zeta)} \iota(u, z\,;\, v, \zeta) \in \mathbb{Z}.$$

2.1 Positivity of Intersections and the Adjunction Formula 27

The choice of notation reflects the fact that $[u] \cdot [v]$ turns out to depend only on the homology classes $[u], [v] \in H_2(M)$; in fact, it defines a nondegenerate bilinear symmetric form

$$H_2(M) \otimes H_2(M) \to \mathbb{Z} \colon [u] \otimes [v] \mapsto [u] \cdot [v].$$

More details on this may be found, e.g., in [Bre93].

If u and v have an isolated but nontransverse intersection at $u(z) = v(\zeta) = p$, one can still define a local intersection index $\iota(u, z ; v, \zeta) \in \mathbb{Z}$ as follows: by assumption, z, and ζ each lie in the interiors of smoothly embedded closed disks $\mathcal{D}_z \subset \Sigma$ and $\mathcal{D}_\zeta \subset \Sigma'$, respectively, such that

$$u(\mathcal{D}_z \setminus \{z\}) \cap v(\mathcal{D}_\zeta \setminus \{\zeta\}) = \varnothing.$$

Then one can find a C^∞-small perturbation u_ϵ of u such that $u_\epsilon|_{\mathcal{D}_z} \pitchfork v|_{\mathcal{D}_\zeta}$ but $u_\epsilon(\partial \mathcal{D}_z)$ and $v(\partial \mathcal{D}_\zeta)$ remain disjoint. We set

$$\iota(u, z ; v, \zeta) := \sum_{u_\epsilon(z')=v(\zeta')} \iota(u_\epsilon, z' ; v, \zeta') \in \mathbb{Z},$$

where the sum is restricted to pairs $(z', \zeta') \in \mathcal{D}_z \times \mathcal{D}_\zeta$.

Exercise 2.1 Suppose Σ and Σ' are compact oriented surfaces with boundary, M is a smooth oriented 4-manifold and

$$f_\tau \colon \Sigma \to M, \qquad g_\tau \colon \Sigma' \to M, \qquad \tau \in [0, 1]$$

are homotopies[1] of maps with the property that for all $\tau \in [0, 1]$,

$$f_\tau(\partial \Sigma) \cap g_\tau(\Sigma') = f_\tau(\Sigma) \cap g_\tau(\partial \Sigma') = \varnothing.$$

Show that if f_τ and g_τ are of class C^1 and have only transverse intersections for $\tau \in \{0, 1\}$, then

$$\sum_{f_0(z)=g_0(\zeta)} \iota(f_0, z ; g_0, \zeta) = \sum_{f_1(z)=g_1(\zeta)} \iota(f_1, z ; g_1, \zeta). \tag{2.2}$$

Deduce from this that the above definition of the local intersection index for an isolated but nontransverse intersection is well defined and independent of the choice of perturbation. Then, show that (2.2) also holds if the intersections for $\tau \in \{0, 1\}$ are assumed to be isolated but not necessarily transverse. *Hint:* If you have never read [Mil97], you should.

---

[1] We are not specifying the regularity of the homotopy in this statement because it does not matter: one can use general perturbation results as in [Hir94] to replace any continuous homotopy between two C^1-smooth maps with a homotopy of class C^1. If desired, one can also perturb all of the maps to make them smooth.

28 Intersections, Ruled Surfaces, and Contact Boundaries

The following useful result is immediate from the above definition; it can be paraphrased by saying that "algebraically nontrivial intersections cannot be perturbed away."

Proposition 2.2 *If* $u: \Sigma \to M$ *and* $v: \Sigma' \to M$ *have an isolated intersection* $u(z) = v(\zeta)$ *with* $\iota(u, z\,;\, v, \zeta) \neq 0$, *then for any neighborhood* $z \in \mathcal{U}_z \subset \Sigma$, *any sufficiently* C^0-*close perturbation* u_ϵ *of* u *satisfies* $u(\mathcal{U}_\epsilon) \cap v(\Sigma') \neq \varnothing$. $\qquad\square$

Recall next that any complex structure on a real vector space induces a preferred orientation. In the case where $u: (\Sigma, j) \to (M, J)$ and $v: (\Sigma', j') \to (M, J)$ are both J-holomorphic curves, this means that each space in (2.1) carries a canonical orientation and they are automatically compatible with the direct sum; hence $\iota(u, z\,;\, v, \zeta) = +1$. This positivity phenomenon turns out to be true for nontransverse intersections as well:

Theorem 2.3 (*local positivity of intersections*) *Suppose* $u: (\Sigma, j) \to (M, J)$ *and* $v: (\Sigma', j') \to (M, J)$ *are nonconstant pseudoholomorphic maps with* $u(z) = v(\zeta) = p \in M$ *for some* $z \in \Sigma$, $\zeta \in \Sigma'$. *Then there exist neighborhoods* $z \in \mathcal{U}_z \subset \Sigma$ *and* $\zeta \in \mathcal{U}_\zeta \subset \Sigma'$ *such that either* $u(\mathcal{U}_z) = v(\mathcal{U}_\zeta)$ *or*

$$u(\mathcal{U}_z \setminus \{z\}) \cap v(\mathcal{U}_\zeta \setminus \{\zeta\}) = \varnothing.$$

Moreover, in the latter case, if $\dim M = 4$, *then* $\iota(u, z\,;\, v, \zeta) \geq 1$, *with equality if and only if the intersection is transverse.*

A proof of this theorem is given in Appendix B.

To understand the global consequences of Theorem 2.3, observe that there are certain obvious situations where a pair of closed J-holomorphic curves $u: (\Sigma, j) \to (M, J)$ and $v: (\Sigma', j') \to (M, J)$ have infinitely many intersections, e.g., if they represent the same curve up to parametrization or they are multiple covers of the same simple curve. In such cases, u and v have globally identical images, and we find neighborhoods with $u(\mathcal{U}_z) = v(\mathcal{U}_\zeta)$ in Theorem 2.3. One can show that in all other cases the set of intersections is finite, a phenomenon known as **unique continuation**. Theorem 2.3 then implies the following corollary:

Corollary 2.4 (*global positivity of intersections*) *If* $\dim M = 4$ *and* $u: (\Sigma, j) \to (M, J)$ *and* $v: (\Sigma', j') \to (M, J)$ *are closed connected J-holomorphic curves with nonidentical images, then they have finitely many intersections, and*

$$[u] \cdot [v] \geq \#\left\{ (z, \zeta) \in \Sigma \times \Sigma' \mid u(z) = v(\zeta) \right\},$$

with equality if and only if all the intersections are transverse. In particular, $[u] \cdot [v] = 0$ *if and only if* $u(\Sigma) \cap v(\Sigma') = \varnothing$. $\qquad\square$

2.1 Positivity of Intersections and the Adjunction Formula 29

We next consider the question of how many times a single closed J-holomorphic curve $u: (\Sigma, j) \to (M, J)$ intersects itself at two distinct points in its domain, i.e., its count of **double points**. This question obviously has no reasonable answer if u is multiply covered, so let us assume u is simple, in which case it has only finitely many double points. We say that a point $z \in \Sigma$ is a **critical point** of u if

$$du(z) = 0.$$

Remark 2.5 This usage of the term "critical point" conflicts with standard terminology, since typically $\dim \Sigma < \dim M$; hence $du(z)$ can never be surjective and u therefore cannot have any regular points, strictly speaking. Note, however, that whenever $du(z) \neq 0$, the Cauchy–Riemann equation implies that $du(z)$ is injective. For this reason, we will refer to points with this property as **immersed points** instead of "regular points."

A simple J-holomorphic curve can have critical points, but only finitely many, and their role in intersection theory is dictated by the following lemma. For an oriented surface Σ and a symplectic manifold (M, ω), we say that a smooth map $u: \Sigma \to M$ is **symplectically immersed** if $u^*\omega > 0$.

Lemma 2.6 *If $u \in \mathcal{M}_g(M, J)$ is simple, then for any parametrization $u: \Sigma \to M$ and any $z \in \Sigma$, there is a neighborhood $z \in \mathcal{U}_z \subset \Sigma$ such that $u|_{\mathcal{U}_z}$ is injective. Moreover, if $du(z) = 0$, $\dim M = 4$, and ω_z is an auxiliary choice of symplectic form defined near $u(z)$ and taming J, then there exists a positive integer $\delta(u, z) > 0$ depending only on the germ of u near z such that $u|_{\mathcal{U}_z}$ admits a C^1-small perturbation to an ω_z-symplectically immersed map $u_\epsilon: \mathcal{U}_z \to M$ that matches u outside an arbitrarily small neighborhood of z and satisfies[2]*

$$\delta(u, z) = \frac{1}{2} \sum_{u_\epsilon(\zeta_1) = u_\epsilon(\zeta_2),\, \zeta_1 \neq \zeta_2} \iota(u_\epsilon, \zeta_1 \,;\, u_\epsilon, \zeta_2),$$

where the sum is finite and ranges over pairs $(\zeta_1, \zeta_2) \in \mathcal{U}_z \times \mathcal{U}_z$.

A proof of this lemma is given in Appendix B. It enables us to define for each simple curve $u \in \mathcal{M}_g(M, J)$ the integer

$$\delta(u) := \frac{1}{2} \sum_{u(z) = u(\zeta),\, z \neq \zeta} \iota(u, z \,;\, u, \zeta) + \sum_{du(z)=0} \delta(u, z) \in \mathbb{Z}, \tag{2.3}$$

---

[2] Notice that each geometric double point $u_\epsilon(\zeta_1) = u_\epsilon(\zeta_2)$ appears twice in the summation over pairs (ζ_1, ζ_2); hence the factor of $1/2$ in the definition of $\delta(u, z)$ and similarly in (2.3).

30 *Intersections, Ruled Surfaces, and Contact Boundaries*

which we shall call the **singularity index** of u. The contribution $\delta(u,z) > 0$ for each critical point z is the **local singularity index** of u at z.

Theorem 2.7 *For any simple curve $u \in M_g(M,J)$ in an almost complex 4-manifold (M,J), the integer $\delta(u)$ defined in (2.3) depends only on the genus g and the homology class $[u] \in H_2(M)$. Moreover, $\delta(u) \geq 0$, with equality if and only if u is embedded.*

Note that the second statement in Theorem 2.7 is an immediate consequence of Theorem 2.3 and Lemma 2.6. To prove the first statement, we shall relate $\delta(u)$ to other quantities that more obviously depend only on $[u] \in H_2(M)$ and the genus, for instance the homological self-intersection number

$$[u] \cdot [u] \in \mathbb{Z}.$$

To compute the latter, it suffices to compute $[u_\epsilon] \cdot [u_\epsilon]$ for any C^1-small immersed perturbation $u_\epsilon : \Sigma \to M$ of u. Choose u_ϵ to be the perturbation promised by Lemma 2.6, so for some auxiliary symplectic structure ω taming J near the images of the critical points of u, we can assume u_ϵ is symplectically immersed near those critical points and matches u everywhere else. Notice that by Lemma 2.6 and the definition of $\delta(u)$,

$$\delta(u_\epsilon) = \delta(u).$$

Denote the normal bundle of u_ϵ by $N_{u_\epsilon} \to \Sigma$. Since u_ϵ is symplectically immersed in the region where it differs from u, we can deform the natural complex structure of $u_\epsilon^* TM$ on this region to one that is tamed by ω but also admits a splitting of complex vector bundles $u_\epsilon^* TM \cong T\Sigma \oplus N_{u_\epsilon}$. This modification of the complex structure does not change $c_1(u_\epsilon^* TM)$, so we then have

$$c_1([u]) = c_1(u_\epsilon^* TM) = c_1(T\Sigma) + c_1(N_{u_\epsilon}) = \chi(\Sigma) + c_1(N_{u_\epsilon}). \tag{2.4}$$

This motivates the following notion: we define the **normal Chern number** $c_N(u) \in \mathbb{Z}$ of any closed J-holomorphic curve $u : (\Sigma, j) \to (M,J)$ to be

$$c_N(u) := c_1([u]) - \chi(\Sigma). \tag{2.5}$$

It is equal to $c_1(N_u)$ whenever u is immersed, but it has the advantage of obviously depending only on $[u] \in H_2(M)$ and the topology of the domain, so we can define it without assuming that u is immersed.

The self-intersection number $[u] \cdot [u] = [u_\epsilon] \cdot [u_\epsilon]$ can now be computed by counting (with signs) the isolated intersections between u_ϵ and a generic perturbation of the form

$$u_\epsilon' : \Sigma \to M : z \mapsto \exp_{u_\epsilon(z)} \eta(z),$$

2.1 Positivity of Intersections and the Adjunction Formula 31

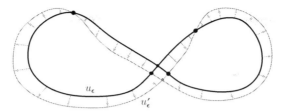

Figure 2.1 Counting the intersections of $u_\epsilon \colon \Sigma \to M$ with a perturbation of the form $u'_\epsilon = \exp_{u_\epsilon} \eta$ for some section η of the normal bundle.

where η is a generic C^0-small smooth section of $N_{u_\epsilon} \to \Sigma$ and the exponential map is defined using any choice of Riemannian metric on M. Figure 2.1 shows how many intersections we should expect to see. Any zero of η with order $k \in \mathbb{Z}$ will produce an intersection of u_ϵ and u'_ϵ whose local intersection index is also k, and the sum of these orders over all zeroes of η is $c_1(N_{u_\epsilon})$. Moreover, any isolated double point $u_\epsilon(z) = u_\epsilon(\zeta)$ will produce *two* intersections of u_ϵ and u'_ϵ with the same local index. These two observations produce the formula

$$[u] \cdot [u] = 2\delta(u_\epsilon) + c_1(N_{u_\epsilon}) = 2\delta(u) + c_N(u).$$

Since neither $[u] \cdot [u]$ nor $c_N(u)$ depends on the perturbation u_ϵ, this proves the following important result, known as the **adjunction formula**, which implies Theorem 2.7 as an immediate corollary:

Theorem 2.8 (*adjunction formula*) *For any closed, connected and simple J-holomorphic curve u in an almost complex 4-manifold (M, J),*

$$[u] \cdot [u] = 2\delta(u) + c_N(u), \tag{2.6}$$

where $c_N(u) \in \mathbb{Z}$ is the normal Chern number (2.5) *and $\delta(u)$ is a nonnegative integer that vanishes if and only if u is embedded.* □

Corollary 2.9 *If $u \in M_g^A(M, J)$ is embedded, then every other simple curve in $M_g^A(M, J)$ is also embedded.* □

Exercise 2.10

(a) Consider the intersecting holomorphic maps $u, v \colon \mathbb{C} \to \mathbb{C}^2$ defined by

$$u(z) = (z^3, z^5), \qquad v(z) = (z^4, z^6).$$

Show that u admits a C^1-small perturbation to a holomorphic function u_ϵ such that u_ϵ and v have exactly 18 intersections in a neighborhood of the origin, all transverse.

32   *Intersections, Ruled Surfaces, and Contact Boundaries*

(b) Try to convince yourself that the above count of 18 intersections holds after *any* generic C^1-small perturbation of u and/or v.

(c) Show that for any neighborhood $\mathcal{U} \subset \mathbb{C}$ of 0, the map u admits a C^1-small perturbation to a holomorphic *immersion* u_ϵ such that

$$\frac{1}{2} \#\{(z, \zeta) \in \mathcal{U} \times \mathcal{U} \mid u_\epsilon(z) = u_\epsilon(\zeta), \ z \neq \zeta\} = 10.$$

(d) If you're especially ambitious, now try to convince yourself that for *any* perturbation as in part (c) making all double points of u_ϵ transverse, the count of double points is the same.

Exercise 2.11 Recall that $H_2(\mathbb{CP}^2)$ is generated by an embedded sphere $\mathbb{CP}^1 \subset \mathbb{CP}^2$ with $[\mathbb{CP}^1] \cdot [\mathbb{CP}^1] = 1$. A holomorphic curve $u \colon \Sigma \to \mathbb{CP}^2$ is said to have **degree** $d \in \mathbb{N}$ if

$$[u] = d[\mathbb{CP}^1].$$

Show that all holomorphic spheres of degree 1 are embedded and any other simple holomorphic sphere in $\mathbb{CP}^2$ is embedded if and only if it has degree 2.

2.2 Application to Ruled Surfaces

We now apply the results of the previous section to complete the proof of Lemma 1.18 from Lecture 1.

Since $\mathcal{M}_0^{[S]}(M, J)$ contains the embedded curve u_S by construction, Corollary 2.9 implies that all other *simple* curves in $\mathcal{M}_0^{[S]}(M, J)$ are also embedded, and we saw in §1.3 that every embedded curve $u \in \mathcal{M}_0^{[S]}(M, J)$ has a neighborhood in $\mathcal{M}_0^{[S]}(M, J)$ consisting of embeddings that foliate an open subset. On a more global level, any two curves $u, v \in \mathcal{M}_0^{[S]}(M, J)$ satisfy

$$[u] \cdot [v] = [S] \cdot [S] = 0,$$

and thus Corollary 2.4 now implies that u and v are disjoint unless they are identical; hence the set of *all* simple curves in $\mathcal{M}_0^{[S]}(M, J)$ foliates an open subset of M.

We must still rule out the possibility that $\mathcal{M}_0^{[S]}(M, J)$ contains a multiple cover, so arguing by contradiction, suppose $u \in \mathcal{M}_0^{[S]}(M, J)$ is a k-fold cover of a simple curve $v \colon (\Sigma_g, j) \to (M, J)$ with genus $g \geq 0$, for some $k \geq 2$. This requires the existence of a map $\varphi \colon S^2 \to \Sigma_g$ of degree k, but such a map cannot exist if $g > 0$ since Σ_g then has a contractible universal cover, and thus $\pi_2(\Sigma_g) = 0$; we conclude $g = 0$. Moreover, the fact that the embedded sphere $S \subset M$ has trivial normal bundle implies via the usual splitting $TM|_S = TS \oplus N_S$ that

2.2 Application to Ruled Surfaces

$$c_1([S]) = c_1(TS) + c_1(N_S) = \chi(S) = 2,$$

so $[S] = [u] = k[v]$ implies $2 = kc_1([v])$, and thus $k = 2$ and $c_1([v]) = 1$. Consider now the adjunction formula (2.6) applied to the simple curve v:

$$[v] \cdot [v] = 2\delta(v) + c_N(v) = 2\delta(v) + c_1([v]) - 2.$$

The right-hand side is an odd integer since $c_1([v]) = 1$. However, the left-hand side is 0, as $0 = [S] \cdot [S] = [u] \cdot [u] = k^2[v] \cdot [v]$, so we have a contradiction.

Next, suppose $u_k \in \mathcal{M}_0^{[S]}(M, J)$ is a sequence degenerating to a nodal curve $\{v_+, v_-\} \in \overline{\mathcal{M}}_0^{[S]}(M, J)$, for which Lemma 1.17 guarantees that both v_+ and v_- are simple and satisfy $c_1([v_\pm]) > 0$. Since $[S] = [u_k] = [v_+] + [v_-]$ and $c_1([S]) = 2$, this implies

$$c_1([v_+]) = c_1([v_-]) = 1. \tag{2.7}$$

Since every curve $u \in \mathcal{M}_0^{[S]}(M, J)$ has $c_1([u]) = c_1([S]) = 2$ and is simple, this implies that u and $v_\pm$ can never have identical images, so $[u] \cdot [v_\pm] \geq 0$ by positivity of intersections (Corollary 2.4). Moreover,

$$0 = [S] \cdot [S] = [u] \cdot ([v_+] + [v_-]) = [u] \cdot [v_+] + [u] \cdot [v_-],$$

where both terms at the right are nonnegative; thus both vanish and we conclude via Corollary 2.4 that u is disjoint from both v_+ and v_-.

We claim next that v_+ and v_- cannot be the *same* curve (up to parametrization): indeed, if they are, then we have $[S] = 2[v_+]$, and applying the adjunction formula to v_+ yields the same numerical contradiction as in the case of a multiple cover in $\mathcal{M}_0^{[S]}(M, J)$. It follows now by Corollary 2.4 that v_+ and v_- have finitely many intersections, all of which count positively, and, in fact,

$$[v_+] \cdot [v_-] \geq 1, \tag{2.8}$$

since they must have at least one intersection, namely at the node. Using $[S] = [v_+] + [v_-]$ and (2.7), and plugging in the adjunction formula and (2.7) to compute $[v_\pm] \cdot [v_\pm]$, we find

$$\begin{aligned}
0 = [S] \cdot [S] &= ([v_+] + [v_-]) \cdot ([v_+] + [v_-]) = [v_+] \cdot [v_+] + [v_-] \cdot [v_-] \\
&\quad + 2[v_+] \cdot [v_-] \\
&= 2\delta(v_+) + c_N(v_+) + 2\delta(v_-) + c_N(v_-) + 2[v_+] \cdot [v_-] \\
&= 2\delta(v_+) + 2\delta(v_-) + c_1([v_+]) - \chi(S^2) + c_1([v_-]) - \chi(S^2) \\
&\quad + 2[v_+] \cdot [v_-] \\
&= 2\delta(v_+) + 2\delta(v_-) + 2\left([v_+] \cdot [v_-] - 1\right).
\end{aligned}$$

34 *Intersections, Ruled Surfaces, and Contact Boundaries*

By (2.8), every term in this last sum is nonnegative, implying

$$\delta(v_+) = \delta(v_-) = 0 \quad \text{and} \quad [v_+] \cdot [v_-] = 1.$$

Applying Corollary 2.4 and Theorem 2.8, we deduce that $v_\pm$ are each embedded and intersect each other exactly once, transversely. Applying the adjunction formula again to $v_\pm$ with $c_N(v_\pm) = c_1([v_\pm]) - \chi(S^2) = -1$ then gives

$$[v_\pm] \cdot [v_\pm] = 2\delta(v_\pm) + c_N(v_\pm) = 0 - 1 = -1,$$

so both are J-holomorphic parametrizations of exceptional spheres.

Finally, we show that if $\{v_+, v_-\}$ and $\{v'_+, v'_-\}$ are two nonidentical nodal curves arising as limits of curves in $\mathcal{M}_0^{[S]}(M, J)$, then they are disjoint. Here, "nonidentical" can be taken to mean without loss of generality (i.e., by reversing the labels of v_+ and v_- if necessary) that v_+ is not equivalent to either v'_+ or v'_- up to parametrization, so positivity of intersections gives $[v_+] \cdot [v'_\pm] \geq 0$. It could still happen in theory that v_- is equivalent to one of v'_+ or v'_-, say, for example, the latter, without loss of generality. Then $[v_-] \cdot [v'_-] = -1$ by the above computation, while $[v_+] \cdot [v_-] = [v_+] \cdot [v'_-] = 1$ and $[v'_+] \cdot [v'_-] = [v'_+] \cdot [v_-] = 1$, and thus

$$0 = [S] \cdot [S] = ([v_+] + [v_-]) \cdot ([v'_+] + [v'_-]) = [v_+] \cdot [v'_+] + [v_+] \cdot [v'_-]$$
$$+ [v_-] \cdot [v'_+] + [v_-] \cdot [v'_-] \geq 0 + 1 + 1 - 1 = 1,$$

giving a contradiction. The only remaining possibility is that each of $v_\pm$ is not equivalent to each of $v'_\pm$, so their intersections are all positive, and the expansion above implies that they are all zero; thus both curves in $\{v_+, v_-\}$ are disjoint from both curves in $\{v'_+, v'_-\}$. The proof of Lemma 1.18 is now complete.

To conclude our discussion of the closed case, let us note which properties of the intersection theory we made essential use of in the above argument:

- The pairing $[u] \cdot [v]$ is *homotopy invariant*.
- The condition $[u] \cdot [v] = 0$ guarantees that two curves u and v with nonidentical images are *disjoint*; moreover, if they have a known intersection, then $[u] \cdot [v] = 1$ guarantees that that intersection is *transverse*.
- There is a homotopy invariant number $\delta(u) \geq 0$ defined for simple curves u, which can be computed in terms of $[u] \cdot [u]$ and whose vanishing guarantees that u is *embedded*.

In order to produce a useful theory for studying contact 3-manifolds, we will want the intersection theory defined in the next two lectures for *punctured* holomorphic curves to have all of these same properties.

2.3 Contact Manifolds, Symplectic Fillings and Cobordisms

The goal for the remainder of these lectures will be to explain a generalization of the intersection theory described above that has applications in 3-dimensional contact topology. One way to motivate the study of contact manifolds is by considering symplectic manifolds with boundary.

A vector field V on a symplectic manifold (M, ω) is called a **Liouville vector field** if it satisfies

$$\mathcal{L}_V \omega = \omega;$$

i.e., its flow rescales the symplectic form exponentially. By Cartan's formula for the Lie derivative, this is equivalent to the condition

$$d\lambda = \omega, \quad \text{where } \lambda := \iota_V \omega,$$

and the primitive λ is then called a **Liouville form**. We say in this case that λ is ω-**dual** to V.

Definition 2.12 Suppose (W, ω) is a symplectic manifold with boundary. A boundary component $M \subset \partial W$ is called **convex/concave** if a neighborhood of M admits a Liouville vector field that points transversely outward/inward, respectively, at M.

Exercise 2.13 Suppose M is an oriented hypersurface in a $2n$-dimensional symplectic manifold (W, ω), and V is a Liouville vector field defined near M, with ω-dual Liouville form λ. Show that V is positively/negatively transverse to M if and only if the restriction of $\lambda \wedge (d\lambda)^{n-1}$ to M is a positive/negative volume form, respectively.

Exercise 2.14 Show that in the situation of Exercise 2.13, the spaces of Liouville forms λ defined near $M \subset (W, \omega)$ such that $\lambda \wedge d\lambda^{n-1}|_{TM}$ is a positive or negative volume form are convex.

Exercise 2.13 leads directly to the notion of a contact manifold: we say that a 1-form α on an oriented $(2n - 1)$-dimensional manifold is a (positive) **contact form** if

$$\alpha \wedge (d\alpha)^{n-1} > 0, \tag{2.9}$$

and a (positive, co-oriented) **contact structure** is any smooth co-oriented hyperplane distribution $\xi \subset TM$ that can be defined by $\xi = \ker \alpha$ for some contact form α. Exercises 2.13 and 2.14 show that whenever $M \subset \partial W$ is a convex/concave boundary component of a symplectic manifold (W, ω), the

36 *Intersections, Ruled Surfaces, and Contact Boundaries*

oriented manifold $\pm M$ inherits a positive[3] contact structure, which is unique up to deformation through families of contact structures. Whenever M is closed, Gray's stability theorem (see, e.g., [Gei08]) then implies that the induced contact structure on M is in fact canonical up to *isotopy*.[4]

Exercise 2.15 Show that up to issues of orientation, the contact condition (2.9) is equivalent to the condition that α is nowhere zero and $d\alpha$ restricts to a nondegenerate 2-form on $\xi := \ker \alpha$; i.e., it makes $(\xi, d\alpha) \to M$ a symplectic vector bundle.

Definition 2.16 Given two closed contact manifolds (M_+, ξ_+) and (M_-, ξ_-) of the same dimension, a **symplectic cobordism** from[5] (M_-, ξ_-) to (M_+, ξ_+) is a compact symplectic manifold (W, ω) with

$$\partial W = -M_- \sqcup M_+$$

such that a neighborhood of ∂W admits a Liouville form λ with

$$\ker\left(\lambda|_{TM_\pm}\right) = \xi_\pm.$$

If $M_- = \varnothing$, we call (W, ω) a (strong) **symplectic filling** of (M_+, ξ_+), and if $M_+ = \varnothing$, we say (W, ω) is a **symplectic cap** for (M_-, ξ_-).

There are many interesting questions one can ask about contact manifolds and the existence of symplectic fillings or cobordisms. The strongest results in this area are typically specific to dimension 3, though there has also been considerable recent progress in higher dimensions. Here is a brief sampling of known results:

---

[3] We are assuming M carries its canonical orientation as a boundary component of the symplectic manifold (W, ω), but also using the notation $-M$ to mean the same manifold with reversed orientation – thus a positive contact structure on $-M$ is in fact a negative contact structure on M.

[4] Gray's stability theorem states that any smooth 1-parameter family of contact structures on a closed manifold arises from an isotopy. It is specifically true for contact *structures* and not contact *forms*, and this is one good reason that we regard the contact structure on a convex/concave boundary of a symplectic manifold as a well-defined object, whereas the contact *form* is only auxiliary data.

[5] Certain orientation conventions are not universally agreed upon: there is a vocal minority of authors who would describe what we are defining here as a "symplectic cobordism *from* (M_+, ξ_+) *to* (M_-, ξ_-)." Whichever convention one prefers, one must be consistent about it – unlike topological cobordisms, the existence of a symplectic cobordism in one direction does not imply that one in the other direction also exists!

2.3 Contact Manifolds, Symplectic Fillings and Cobordisms   37

(1) Martinet [Mar71] proved that every closed oriented 3-manifold admits a contact structure. This result was recently extended to all dimensions by Borman, Eliashberg, and Murphy [BEM15], given the obviously necessary topological condition that an *almost* contact structure exists.

(2) A combination of results due to Gromov and Eliashberg [Gro85, Eli90, Eli89] implies that any contact structure on any closed 3-manifold M is homotopic through oriented 2-plane fields to a contact structure ξ for which (M, ξ) admits no symplectic filling. These are the so-called *overtwisted* contact structures. This notion has also recently been generalized to all dimensions in [BEM15].

(3) A result of Lisca [Lis98] even gives examples of closed oriented 3-manifolds on which *no* contact structure is symplectically fillable. Etnyre and Honda [EH01] later extended this to find 3-manifolds on which every contact structure is overtwisted.

(4) Etnyre and Honda [EH02] showed that, in contrast to fillings, symplectic *caps* do exist for any closed contact 3-manifold, and, in fact, they come in infinitely many distinct topological types. The existence of caps in all higher dimensions was established only very recently, in the parallel work of Conway and Etnyre [CE] and Lazarev [Laz].

(5) Etnyre and Honda [EH02] also showed that every closed overtwisted contact 3-manifold admits a symplectic cobordism to every other closed contact 3-manifold. The higher-dimensional analogue of this result was recently established by Eliashberg and Murphy [EM].

Let us state more carefully two further results along these lines that will be discussed in Lecture 5. We say that two symplectic manifolds (W, ω) and (W', ω') with convex boundary are **symplectically deformation equivalent** if there is a diffeomorphism $\varphi \colon W \to W'$ such that $\varphi^* \omega'$ can be deformed to ω through a smooth 1-parameter family of symplectic forms that are all convex at the boundary. The **standard contact structure** ξ_{std} on S^3 is defined by identifying S^3 with the boundary of the unit ball B^4 with its **standard symplectic form on the 4-ball**

$$\omega_{\mathrm{std}} := \sum_{j=1}^{2} dx_j \wedge dy_j$$

and Liouville form

$$\lambda_{\mathrm{std}} := \frac{1}{2} \sum_{j=1}^{2} \left(x_j \, dy_j - y_j \, dx_j \right).$$

By this definition, $(B^4, \omega_{\mathrm{std}})$ is a symplectic filling of $(S^3, \xi_{\mathrm{std}})$, and one can trivially produce other fillings of $(S^3, \xi_{\mathrm{std}})$ with different topological types by

38   *Intersections, Ruled Surfaces, and Contact Boundaries*

blowing up $(B^4, \omega_{\text{std}})$ in its interior. This procedure, however, produces a fairly limited range of topological types for manifolds W with $\partial W = S^3$. Note that in terms of smooth topology, *almost anything* can have boundary S^3: just take any closed oriented 4-manifold, remove a ball and reverse the orientation. Symplectically, however, the situation is very different:

Theorem 2.17 (Gromov [Gro85])   *Every symplectic filling of (S^3, ξ_{std}) is symplectically deformation equivalent to a blowup of $(B^4, \omega_{\text{std}})$.*

Similarly, $S^1 \times S^2$ and the lens spaces $L(k, k-1)$ for $k \in \mathbb{N}$ each carry standard contact structures as convex boundaries of certain symplectic manifolds, and their fillings are also unique in the sense of the theorem above.

Theorem 2.18   *The contact manifolds $(S^1 \times S^2, \xi_{\text{std}})$ and $(L(k, k-1), \xi_{\text{std}})$ for $k \in \mathbb{N}$ each have unique symplectic fillings up to deformation equivalence and blowup.*

Theorem 2.18 was proved for $S^1 \times S^2$ originally by Eliashberg [Eli90], and the uniqueness for $L(k, k-1)$ up to diffeomorphism was proved by Lisca [Lis08]. In the forms stated above, Theorems 2.17 and 2.18 are both easy applications of a more general result from [Wen10b], which can be thought of as an analogue of McDuff's Theorem 1.16 for symplectic fillings of certain contact 3-manifolds. This will be the main subject of Lecture 5.

2.4 Asymptotically Cylindrical Holomorphic Curves

It is not usually useful to consider *closed* holomorphic curves in symplectic cobordisms – for example, the symplectic form on a cobordism could be exact, in which case Stokes's theorem implies that all closed holomorphic curves for a tame almost complex structure are constant. A useful alternative is to consider noncompact holomorphic curves with cylindrical ends, and the proper setting for this is the noncompact *completion* of a symplectic cobordism. The study of holomorphic curves in this setting is a large subject known as *symplectic field theory* (see [EGH00, Wenb]), and we shall only touch upon a few aspects of it here.

Assume (W, ω) is a symplectic cobordism from (M_-, ξ_-) to (M_+, ξ_+), with a neighborhood of ∂W admitting a Liouville form λ such that

$$\xi_\pm = \ker \alpha_\pm, \quad \text{where } \alpha_\pm := \lambda|_{TM_\pm}.$$

Exercise 2.19   Show that the flow from $M_\pm$ along the Liouville vector field dual to λ identifies collar neighborhoods $\mathcal{N}(M_\pm) \subset W$ of $M_\pm$ with the models

2.4 Asymptotically Cylindrical Holomorphic Curves

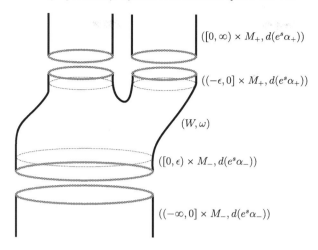

Figure 2.2 The completion of a symplectic cobordism is constructed by attaching half-symplectizations to form cylindrical ends.

$$(\mathcal{N}(M_+), \lambda) \cong ((-\epsilon, 0] \times M_+, e^s \alpha_+),$$
$$(\mathcal{N}(M_-), \lambda) \cong ([0, \epsilon) \times M_-, e^s \alpha_-)$$

for sufficiently small $\epsilon > 0$, where s denotes the real coordinate in $(-\epsilon, 0]$ or $[0, \epsilon)$.

For any contact manifold $(M, \xi = \ker \alpha)$, the exact symplectic manifold $(\mathbb{R} \times M, d(e^s \alpha))$ is called the **symplectization** of (M, ξ); one can show that its symplectomorphism type depends on ξ but not on the choice of contact form α. A choice of α does, however, determine a distinguished vector field that spans the characteristic line fields of the hypersurfaces $\{s\} \times M$: we define the **Reeb vector field** to be the unique vector field R_α on M satisfying

$$d\alpha(R_\alpha, \cdot) \equiv 0 \quad \text{and} \quad \alpha(R_\alpha) \equiv 1.$$

The **symplectic completion** of the cobordism (W, ω) is defined by attaching halves of symplectizations along the collar neighborhoods from Exercise 2.19, producing the noncompact symplectic manifold (see Figure 2.2).

$$(\widehat{W}, \widehat{\omega}) := ((-\infty, 0] \times M_-, d(e^s \alpha_-)) \cup_{M_-} (W, \omega)$$
$$\cup_{M_+} ([0, \infty) \times M_+, d(e^s \alpha_+)). \qquad (2.10)$$

Informally, the symplectization of (M, ξ) can also be thought of as the completion of a trivial symplectic cobordism from (M, ξ) to itself.

40   *Intersections, Ruled Surfaces, and Contact Boundaries*

Given a choice of contact form α for ξ, $(\mathbb{R} \times M, d(e^s \alpha))$ carries a special class $\mathcal{J}(\alpha)$ of compatible almost complex structures J, defined by the following conditions:

- $J(\partial_s) = R_\alpha$.
- $J(\xi) = \xi$ and $J|_\xi$ is compatible with $d\alpha|_\xi$.
- J is invariant under the translation action $(s, p) \mapsto (s + c, p)$ for all $c \in \mathbb{R}$.

For any $J \in \mathcal{J}(\alpha)$, a periodic orbit $x \colon \mathbb{R} \to M$ of R_α with period $T > 0$ gives rise to a J-holomorphic cylinder

$$ u \colon \mathbb{R} \times S^1 \to \mathbb{R} \times M \colon (s, t) \mapsto (Ts, x(Tt)). $$

Such curves are referred to as **orbit cylinders** (sometimes also **trivial cylinders**), and they serve as asymptotic models for the more general class of holomorphic curves that we now wish to consider.

Indeed, on the completion $(\widehat{W}, \widehat{\omega})$ as defined in (2.10), let $\mathcal{J}(\omega, \alpha_+, \alpha_-)$ denote the space of almost complex structures that are compatible with ω on W and belong to $\mathcal{J}(\alpha_\pm)$ on $[0, \infty) \times M_+$ and $(-\infty, 0] \times M_-$, respectively. A choice of $J \in \mathcal{J}(\omega, \alpha_+, \alpha_-)$ makes $(\widehat{W}, J)$ into an **almost complex manifold with cylindrical ends**. A Riemann surface with cylindrical ends can likewise be constructed by introducing *punctures* into a closed Riemann surface. That is, suppose (Σ, j) is closed and $\Gamma \subset \Sigma$ is a finite set partitioned into two subsets, $\Gamma = \Gamma^+ \sqcup \Gamma^-$, which we will call the **positive** and **negative punctures**, writing the resulting punctured surface as

$$ \dot{\Sigma} := \Sigma \setminus \Gamma. $$

Near each $z \in \Gamma^\pm$, one can identify a closed neighborhood $\mathcal{D}_z \subset \Sigma$ of z biholomorphically with the standard unit disk $(\mathbb{D}, i)$ such that z is identified with the origin, and then identify $\mathbb{D} \setminus \{0\}$ in turn with a half-cylinder via the biholomorphic map

$$ [0, \infty) \times S^1 \to \mathbb{D} \setminus \{0\} \colon (s, t) \mapsto e^{-2\pi(s+it)}, \quad \text{for } z \in \Gamma^+, $$
$$ (-\infty, 0] \times S^1 \to \mathbb{D} \setminus \{0\} \colon (s, t) \mapsto e^{2\pi(s+it)}, \quad \text{for } z \in \Gamma^-. $$

We will refer to this identification as a choice of **cylindrical coordinates** near $z \in \Gamma^\pm$. Making such a choice for all punctures, this determines a decomposition

$$ \dot{\Sigma} = ((-\infty, 0] \times C_-) \cup_{C_-} \Sigma_0 \cup_{C_+} ([0, \infty) \times C_+) \tag{2.11} $$

analogous to (2.10), where $\Sigma_0 := \Sigma \setminus \bigcup_{z \in \Gamma} \mathring{\mathcal{D}}_z$ can be regarded as a cobordism with $\partial \Sigma_0 = -C_- \sqcup C_+$ between two disjoint unions of circles $C_\pm$, and the

2.4 Asymptotically Cylindrical Holomorphic Curves

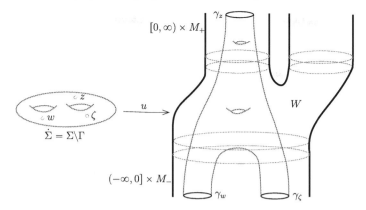

Figure 2.3 An asymptotically cylindrical map $u \colon \dot{\Sigma} \to \widehat{W}$ of a punctured surface of genus 2 into a completed cobordism, with one positive puncture $z \in \Sigma$ asymptotic to a Reeb orbit γ_z in M_+ and two positive punctures $w, \zeta \in \Sigma$ asymptotic to Reeb orbits γ_w and γ_ζ in M_-.

complex structure on the cylindrical ends is always the standard one, i.e., with $i\partial_s = \partial_t$ in cylindrical coordinates (s, t).

We say that a smooth map $u \colon \dot{\Sigma} \to \widehat{W}$ is (positively or negatively) **asymptotic** at $z \in \Gamma^\pm$ to a T-periodic orbit $x \colon \mathbb{R} \to M_\pm$ of $R_{\alpha_\pm}$ if there exists a choice of cylindrical coordinates as above in which u near z takes the form

$$u(s, t) = \exp_{(Ts, x(Tt))} h(s, t) \in \mathbb{R} \times M_\pm, \quad \text{for } |s| \text{ large},$$

where the exponential map is defined with respect to a translation-invariant choice of Riemannian metric on $\mathbb{R} \times M_\pm$ and $h(s, t)$ is a vector field along the orbit cylinder that decays to 0 with all derivatives as $s \to \pm\infty$. We say that $u \colon \dot{\Sigma} \to M$ is **asymptotically cylindrical** if it is positively/negatively asymptotic to some closed Reeb orbit in M_+ or M_-, respectively, at each of its positive/negative punctures (see Figure 2.3).

Observe that the completion $\widehat{W}$ admits a natural compactification as a compact topological manifold with boundary:

$$\overline{W} := ([-\infty, 0] \times M_-) \cup_{M_-} W \cup_{M_+} ([0, \infty] \times M_+).$$

In the same way, the decomposition (2.11) allows us to define the **circle compactification** $\overline{\Sigma}$ of $\dot{\Sigma}$, a compact topological 2-manifold with boundary whose interior is identified with $\dot{\Sigma}$. It follows, then, from the definition above that any asymptotically cylindrical map $u \colon \dot{\Sigma} \to \widehat{W}$ extends naturally to a continuous map

42 Intersections, Ruled Surfaces, and Contact Boundaries

$$\bar{u} \colon \overline{\Sigma} \to \overline{W},$$

which takes each component of $\partial\overline{\Sigma}$ to a closed Reeb orbit in $\{\pm\infty\} \times M_{\pm}$.

If the set of punctures is nonempty, then an asymptotically cylindrical map $u \colon \dot{\Sigma} \to \widehat{W}$ does not represent a homology class in $H_2(\widehat{W})$, but one can use the compactifications described above to assign it a *relative* homology class. Concretely, let

$$\overline{\gamma}^{\pm} \subset \{\pm\infty\} \times M_{\pm} \subset \partial\overline{W}$$

denote the union of all of the images of the positive/negative asymptotic orbits of u; topologically, this is a disjoint union of embedded circles. The **relative homology class** of u is then defined as

$$[u] := \bar{u}_*[\overline{\Sigma}] \in H_2(\overline{W}, \overline{\gamma}^+ \cup \overline{\gamma}^-),$$

where $[\overline{\Sigma}] \in H_2(\overline{\Sigma}, \partial\overline{\Sigma})$ denotes the relative fundamental class of $\overline{\Sigma}$. The long exact sequence of the pair $(\overline{W}, \overline{\gamma}^+ \cup \overline{\gamma}^-)$ implies that any two asymptotically cylindrical maps having the same asymptotic orbits with the same multiplicities have relative homology classes that differ by a unique *absolute* homology class, that is, a class in the image of the natural map $H_2(\overline{W}) \to H_2(\overline{W}, \overline{\gamma}^+ \cup \overline{\gamma}^-)$; note that the latter is injective since $H_2(\overline{\gamma}^+ \cup \overline{\gamma}^-) = 0$. In most situations, it is convenient to apply the obvious deformation retraction $\overline{W} \to W$ and regard $[u]$ as an element of $H_2(W, \overline{\gamma}^+ \cup \overline{\gamma}^-)$, with $\overline{\gamma}^{\pm}$ now regarded as submanifolds of $M_{\pm} \subset \partial W$. In the special case where $(\widehat{W}, \widehat{\omega})$ is just the symplectization of a single contact manifold (M, ξ), we take this one step further and retract $\mathbb{R} \times M$ to $\{0\} \times M$, so that $[u]$ lives naturally in $H_2(M, \overline{\gamma}^+ \cup \overline{\gamma}^-)$. In the following discussion, we will state definitions assuming that $(\widehat{W}, \widehat{\omega})$ is a completion of a nontrivial cobordism instead of a symplectization, but one can make obvious modifications to accommodate the latter case.

Moduli spaces of punctured J-holomorphic curves are now defined as follows: choose finite ordered sets of closed Reeb orbits

$$\gamma^+ = (\gamma_1^+, \ldots, \gamma_{r_+}^+) \text{ in } M_+ \qquad \text{and} \qquad \gamma^- = (\gamma_1^-, \ldots, \gamma_{r_-}^-) \text{ in } M_-,$$

and a relative homology class $A \in H_2(W, \overline{\gamma}^+ \cup \overline{\gamma}^-)$, where $\overline{\gamma}^{\pm}$ denotes the union of the images of the orbits $\gamma_1^{\pm}, \ldots, \gamma_{r_{\pm}}^{\pm}$. We then define

$$\mathcal{M}_g^A(\widehat{W}, J; \gamma^+, \gamma^-) := \{(\Sigma, j, \Gamma^+, \Gamma^-, u)\} \big/ \sim,$$

where

- (Σ, j) is a closed connected Riemann surface of genus g.
- $\Gamma^{\pm} = (z_1^{\pm}, \ldots, z_{r_{\pm}}^{\pm})$ are disjoint finite ordered sets of pairwise distinct points in Σ, defining a punctured surface $\dot{\Sigma} := \Sigma \setminus (\Gamma^+ \cup \Gamma^-)$.

2.4 Asymptotically Cylindrical Holomorphic Curves

43

- The map $u : (\dot{\Sigma}, j) \to (\widehat{W}, J)$ is J-holomorphic, asymptotic to $\gamma_i^{\pm}$ at $z_i^{\pm} \in \Gamma^{\pm}$ for $i = 1, \ldots, r_{\pm}$, and represents the relative homology class A.
- Two such tuples are considered equivalent if they are related by a biholomorphic map that preserves the sets of positive and negative punctures, along with their orderings.

We shall denote unions of these spaces over all possible choices of data by

$$
\mathcal{M}_g(\widehat{W}, J; \gamma^+, \gamma^-) := \bigsqcup_{A \in H_2(W, \bar{\gamma}^+ \cup \bar{\gamma}^-)} \mathcal{M}_g^A(\widehat{W}, J; \gamma^+, \gamma^-),
$$

$$
\mathcal{M}_{g, r_+, r_-}(\widehat{W}, J) := \bigsqcup_{|\gamma^{\pm}| = r_{\pm}} \mathcal{M}_g(\widehat{W}, J; \gamma^+, \gamma^-),
$$

$$
\mathcal{M}_g(\widehat{W}, J) := \bigsqcup_{r_+, r_- \geq 0} \mathcal{M}_{g, r_+, r_-}(\widehat{W}, J).
$$

A topology on $\mathcal{M}_g(\widehat{W}, J)$ can be defined by saying that a sequence $[(\Sigma_k, j_k, \Gamma_k^+, \Gamma_k^-, u_k)]$ converges to an element $[(\Sigma, j, \Gamma^+, \Gamma^-, u)]$ if there exist representatives $(\Sigma, j_k', \Gamma^+, \Gamma^-, u_k') \sim (\Sigma_k, j_k, \Gamma_k^+, \Gamma_k^-, u_k)$ such that

$$
j_k' \to j \text{ in } C^{\infty}(\Sigma), \qquad u_k' \to u \text{ in } C^{\infty}_{\text{loc}}(\dot{\Sigma}, \widehat{W}), \qquad \text{and} \qquad \bar{u}_k' \to \bar{u} \text{ in } C^0(\overline{\Sigma}, \overline{W}).
$$

Our goal for the next pair of lectures will be to write down generalizations of the homological intersection number and the adjunction formula for curves in $\mathcal{M}_g(\widehat{W}, J)$. These will be instrumental in the proof of Theorems 2.17 and 2.18.

Lecture 3

Asymptotics of Punctured Holomorphic Curves

If $u_1 \in \mathcal{M}_g(\widehat{W}, J; \boldsymbol{\gamma}_1^+, \boldsymbol{\gamma}_1^-)$ and $u_2 \in \mathcal{M}_g(\widehat{W}, J; \boldsymbol{\gamma}_2^+, \boldsymbol{\gamma}_2^-)$ are two asymptotically cylindrical holomorphic curves in a 4-dimensional completed symplectic cobordism, it remains true as in the closed case, that intersections of u_1 with u_2 are isolated and positive unless both curves have identical images (i.e., they cover the same simple curve up to parametrization). Since the domains are no longer compact, however, it is not obvious whether the number of intersections is still finite. If it is finite, then one can define an algebraic intersection number

$$u_1 \cdot u_2 \in \mathbb{Z},$$

which is guaranteed to be nonnegative and strictly positive unless the two curves are disjoint. Such a number is not very useful, though, unless it is *homotopy invariant*; i.e., we would like to know that for any family $u_s \in \mathcal{M}_g$ $(\widehat{W}, J; \boldsymbol{\gamma}_1^+, \boldsymbol{\gamma}_1^-)$ that depends continuously (with respect to the topology of the moduli space) on a parameter $s \in [0, 1]$, we have $u_0 \cdot u_2 = u_1 \cdot u_2$. This turns out to be *false* in general, as the noncompactness of the domains can allow intersections to escape to infinity and disappear under homotopies (see Figure 3.1). It is a very powerful fact, first suggested by Hofer and then worked out in detail by Siefring [Sie05, Sie08, Sie11], that this phenomenon can be controlled: one can define for any two distinct punctured holomorphic curves a count of *virtual* intersections that are "hidden at infinity" such that the sum of this number with $u_1 \cdot u_2$ is homotopy invariant. We will define this precisely in the next lecture and explain some applications in Lecture 5. As a preliminary step, it is necessary to gain a fairly precise understanding of the asymptotic behavior of punctured holomorphic curves, so that will be the topic for this lecture.

Remark 3.1   While all results in this and the next lecture are stated in the setting of symplectizations of contact manifolds and (completed) symplectic cobordisms between them, they are valid in somewhat greater generality: they

3.1 Holomorphic Half-Cylinders as Gradient-Flow Lines

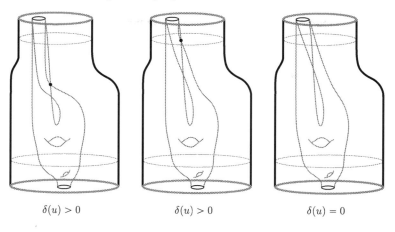

$\delta(u) > 0$ $\delta(u) > 0$ $\delta(u) = 0$

Figure 3.1 The condition of two asymptotically cylindrical holomorphic curves (or two ends of the same curve) being *disjoint* is not homotopy invariant, as intersections can escape to infinity if the two asymptotic Reeb orbits coincide.

continue to hold, namely whenever contact forms are replaced by stable Hamiltonian structures, so long as one can still assume that all closed Reeb orbits are nondegenerate (or Morse–Bott – see the footnote attached to Theorem 4.1). The main results are restated in this more general form in Appendix C.

3.1 Holomorphic Half-Cylinders as Gradient-Flow Lines

Historically, the study of punctured holomorphic curves arose from an analogy with Floer's interpretation of Morse theory as the study of gradient-flow lines of a Morse function (see, e.g., [Sal99, AD14]). In Morse theory, one considers a manifold M with a smooth function $f: M \to \mathbb{R}$, which is called a **Morse function** if its Hessian at every critical point $p \in \mathrm{Crit}(f)$

$$\mathrm{Hess}_p := \nabla df(p) \colon T_pM \times T_pM \to \mathbb{R}$$

is nondegenerate; here ∇ denotes the covariant derivative for any choice of connection on M, but the Hessian does not depend on this choice, since $df(p) = 0$. Recall that the Hessian is automatically a symmetric bilinear map, and if we choose a Riemannian metric g with Levi-Civita connection ∇ and consider instead the covariant derivative of the gradient, we can then identify Hess_p with the linear map

$$A_p := \nabla(\nabla f)(p) \colon T_pM \to T_pM,$$

46 *Asymptotics of Punctured Holomorphic Curves*

which is symmetric with respect to the inner product defined by g. One way of proving the classical Morse inequalities on M is by defining a homology theory with a chain complex generated by critical points in $\text{Crit}(f)$ and a differential defined by counting isolated solutions to the gradient-flow problem

$$\mathcal{M}(p_+, p_-) := \left\{ x \colon \mathbb{R} \to M \;\middle|\; \dot{x} = \nabla f(x) \text{ and } \lim_{s \to \pm\infty} x(s) = p_\pm \right\},$$

for $p_\pm \in \text{Crit}(f)$. In particular, one can show that the resulting homology theory is isomorphic to the usual singular homology $H_*(M)$, thus giving relations between the topology of M and the set of critical points of f (see, e.g., [Sch93, AD14]).

Since the Hessian $A_p = \nabla(\nabla f)(p)$ is symmetric, its eigenvectors in $T_p M$ are orthogonal and its eigenvalues are real. Another way of expressing the Morse condition is to say that $0 \notin \sigma(A_p)$ for all $p \in \text{Crit}(f)$, and the Morse index of p is then the algebraic count of negative eigenvalues in $\sigma(A_p)$. It turns out that the spectrum $\sigma(A_p)$ also controls the asymptotic behavior of gradient-flow lines approaching p: the following result from the theory of ordinary differential equations makes this statement precise:

Proposition 3.2 *Suppose $f \colon M \to \mathbb{R}$ is a Morse function on a Riemannian manifold (M, g), and $x \in \mathcal{M}(p_+, p_-)$ is a gradient-flow line between two critical points $p_+, p_- \in \text{Crit}(f)$. Let $h_\pm(s) \in T_{p_\pm} M$ denote the unique smooth functions defined for s sufficiently close to $\pm\infty$ by*

$$x(s) = \exp_{p_\pm} h_\pm(s).$$

Then there exist unique nontrivial eigenvectors $v_\pm \in T_{p_\pm} M$ of $A_{p_\pm}$ with

$$A_{p_\pm} v_\pm = \lambda_\pm v_\pm, \qquad \lambda_+ < 0 \text{ and } \lambda_- > 0$$

such that $h_+(s)$ and $h_-(s)$ satisfy the exponential decay formula

$$h_\pm(s) = e^{\lambda_\pm s}(v_\pm + r_\pm(s)), \quad \text{for } s \text{ near } \pm\infty,$$

where $r_\pm(s) \in T_{p_\pm} M$ are functions satisfying $r_\pm(s) \to 0$ as $s \to \pm\infty$.

Exercise 3.3 Try to prove the following lemma in the background of Proposition 3.2: suppose S is a real symmetric n-by-n matrix, $A(s)$ is a smooth matrix-valued function with $A(s) \to S$ as $s \to \infty$ and $v(s) \in \mathbb{R}^n$ is a smooth function that is defined for large s, satisfies the linear ODE $\dot{v}(s) - A(s)v(s) = 0$ and decays to 0 as $s \to \infty$. Then $v(s)$ satisfies

$$v(s) = e^{\lambda s}(v_+ + r(s))$$

for a unique eigenvector v_+ of S with $S v_+ = \lambda v_+$ and $\lambda < 0$ and a function $r(s)$ with $r(s) \to 0$ as $s \to \infty$.

3.1 Holomorphic Half-Cylinders as Gradient-Flow Lines 47

One consequence of Proposition 3.2 is that the direction of approach of a gradient-flow line to a nondegenerate critical point is always determined by an eigenvector of the Hessian. We will not discuss this result any further here, but it will serve as motivation for some similar results about asymptotics of J-holomorphic half-cylinders, which can be proved using methods of elliptic regularity theory.

To see what this discussion has to do with holomorphic curves, consider a contact manifold (M, ξ) with contact form α and translation-invariant almost complex structure $J \in \mathcal{J}(\alpha)$ on the symplectization $(\mathbb{R} \times M, d(e^s \alpha))$. Denote the positive/negative half-cylinders by

$$Z_+ := [0, \infty) \times S^1, \qquad Z_- := (-\infty, 0] \times S^1$$

with their standard complex structures defined by $i \partial_s = \partial_t$ in the coordinates (s, t). We defined in §2.4 what it means for a J-holomorphic half-cylinder $u: (Z_\pm, i) \to (\mathbb{R} \times M, J)$ to be asymptotic to a closed Reeb orbit. We claim that such half-cylinders can be regarded in a loose sense as *gradient-flow lines* of a functional on $C^\infty(S^1, M)$ whose critical points are closed Reeb orbits. To see this, let $\pi_\alpha: TM \to \xi$ denote the projection along the Reeb vector field. Then the nonlinear Cauchy–Riemann equation $\partial_s u + J(u) \partial_t u = 0$ satisfied by a map $u = (f, v): Z_\pm \to \mathbb{R} \times M$ is equivalent to the three equations

$$\partial_s f - \alpha(\partial_t v) = 0,$$
$$\partial_t f + \alpha(\partial_s v) = 0, \qquad (3.1)$$
$$\pi_\alpha \partial_s v + J \pi_\alpha \partial_t v = 0.$$

Consider the **contact action functional**

$$\Phi_\alpha: C^\infty(S^1, M) \to \mathbb{R}: \gamma \mapsto \int_{S^1} \gamma^* \alpha.$$

Exercise 3.4 Show that for any smooth 1-parameter family of loops $\gamma_s: S^1 \to M$ with $\gamma := \gamma_0$ and $\eta := \partial_s \gamma_s|_{s=0} \in \Gamma(\gamma^* TM)$,

$$d\Phi_\alpha(\gamma)\eta := \frac{d}{ds}\Phi_\alpha(\gamma_s)\bigg|_{s=0} = \int_{S^1} d\alpha(\eta(t), \dot\gamma(t)) \, dt.$$

Deduce that $\gamma \in C^\infty(S^1, M)$ is a critical point of Φ_α if and only if $\dot\gamma(t) \in \ker d\alpha$ for all t, meaning $\dot\gamma$ is everywhere proportional to R_α.

Observe that Φ_α has a very large symmetry group: it is independent of the choice of parametrization for a loop $\gamma: S^1 \to M$, and, correspondingly, $d\Phi_\alpha(\gamma)\eta$ vanishes for any variation η in the direction of the Reeb vector field. Since the main point of this discussion, however, is to study asymptotic approach to Reeb orbits, we can limit our attention to loops that are C^∞-close

48 *Asymptotics of Punctured Holomorphic Curves*

to Reeb orbits: such loops are always immersions transverse to ξ, and all nearby loops are obtained (up to parametrization) via perturbations along ξ. We shall therefore consider $d\Phi_\alpha(\gamma)$ restricted to sections of $\gamma^*\xi$. Define an L^2-inner product on $\Gamma(\gamma^*\xi)$ by

$$\langle \eta_1, \eta_2 \rangle_{L^2} := \int_{S^1} d\alpha(\eta_1(t), J\eta_2(t))\, dt; \tag{3.2}$$

this is nondegenerate and symmetric since $J|_\xi$ is compatible with $d\alpha|_\xi$. Now for any $\gamma \in C^\infty(S^1, M)$ and $\eta \in \Gamma(\gamma^*\xi)$, we have

$$d\Phi_\alpha(\gamma)\eta = \langle -J\pi_\alpha\dot\gamma, \eta \rangle_{L^2},$$

and thus we can sensibly define $\nabla^\xi\Phi_\alpha(\gamma) := -J\pi_\alpha\dot\gamma$ and interpret the third equation in (3.1) as a gradient flow equation for the family of loops $v(s) := v(s, \cdot) \in C^\infty(S^1, M)$,

$$\pi_\alpha\partial_s v(s) = \nabla^\xi\Phi_\alpha(v(s)). \tag{3.3}$$

This interpretation is mostly formal, as equations like (3.3) typically do not yield well-defined flows on infinite-dimensional Fréchet manifolds such as $C^\infty(S^1, M)$; in reality, one must study these equations as PDEs rather than ODEs and use elliptic theory to obtain results, but the gradient-flow interpretation provides something of a blueprint indicating what results one should try to prove.

For example, it is now reasonable to expect that the asymptotic behavior of solutions to (3.1) might be controlled by the spectrum of some symmetric operator interpreted as the "Hessian" of Φ_α. We deduce the form of this operator as follows: assume $\gamma \colon S^1 \to M$ parametrizes a Reeb orbit with period $T > 0$ such that $\alpha(\dot\gamma(t)) = T$ for all t. Suppose γ_s is a smooth 1-parameter family of loops with $\gamma_0 = \gamma$ and $\partial_s\gamma_s|_{s=0} =: \eta \in \Gamma(\gamma^*\xi)$. Then choosing any symmetric connection ∇ on M, the Hessian of Φ_α at γ should map η to the covariant derivative of $\nabla^\xi\Phi_\alpha$ in the direction η: a computation gives

$$\nabla\left(\nabla^\xi\Phi_\alpha\right)(\gamma)\eta := \nabla_s\left(\nabla^\xi\Phi_\alpha\right)(\gamma_s)\Big|_{s=0} = \nabla_s\left(-J\pi_\alpha\dot\gamma_s\right)|_{s=0}$$
$$= -J(\nabla_t\eta - T\nabla_\eta R_\alpha). \tag{3.4}$$

Note that since $\nabla^\xi\Phi_\alpha(\gamma) = 0$ this expression is independent of the choice of connection. This motivates the following definition:

Definition 3.5 Given a Reeb orbit $\gamma \colon S^1 \to M$ parametrized so that $\alpha(\dot\gamma) \equiv T > 0$ is constant, the **asymptotic operator** associated to γ is

$$\mathbf{A}_\gamma \colon \Gamma(\gamma^*\xi) \to \Gamma(\gamma^*\xi) \colon \eta \mapsto -J(\nabla_t\eta - T\nabla_\eta R_\alpha).$$

Exercise 3.6 Fill in the gaps in the computation (3.4).

3.2 Asymptotic Formulas for Cylindrical Ends 49

Let $H^1(\gamma^*\xi)$ denote the Sobolev space of sections $S^1 \to \gamma^*\xi$ of class L^2 that have weak derivatives also of class L^2. The operator $\mathbf{A}_\gamma$ then extends to a continuous linear map $H^1(\gamma^*\xi) \to L^2(\gamma^*\xi)$. By a similar argument as with the usual Hessian of a smooth function on a finite-dimensional manifold, one can show that $\mathbf{A}_\gamma$ is always symmetric with respect to the L^2-inner product (3.2), and, in fact, the following proposition applies:

Proposition 3.7 ([HWZ95, §3]) *For every Reeb orbit γ, the asymptotic operator $\mathbf{A}_\gamma$ determines an unbounded self-adjoint operator on $L^2(\gamma^*\xi)$ with dense domain $H^1(\gamma^*\xi)$. Its spectrum $\sigma(\mathbf{A}_\gamma)$ consists of real eigenvalues that accumulate at $-\infty$ and $+\infty$ and nowhere else.*

The natural analogue of the Morse condition for Φ_α is now as follows:

Definition 3.8 A Reeb orbit γ is called **nondegenerate** if $\ker \mathbf{A}_\gamma = \{0\}$.

Exercise 3.9 Show that for any contact form α, the flow $\varphi_{R_\alpha}^t$ of the Reeb vector field preserves α for all t, so, in particular, it preserves $\xi = \ker \alpha$ and the symplectic bundle structure $d\alpha|_\xi$. Then show that a Reeb orbit $\gamma \colon S^1 \to M$ of period $T > 0$ is nondegenerate if and only if

$$d\varphi_{R_\alpha}^T|_{\xi_{\gamma(0)}} : \xi_{\gamma(0)} \to \xi_{\gamma(0)}$$

does not have 1 as an eigenvalue. Deduce from this that nondegenerate Reeb orbits are (up to parametrization) always isolated in $C^\infty(S^1, M)$.

3.2 Asymptotic Formulas for Cylindrical Ends

We shall now state some asymptotic results analogous to Proposition 3.2, but for holomorphic curves instead of gradient-flow lines. In the form presented here, these results are due to Siefring [Sie05, Sie08], and they are generalizations and improvements of earlier results of Hofer, Wysocki, and Zehnder [HWZ96a, HWZ96b]; Kriener [Kri98]; and Mora [Mor03]. The proofs are lengthy and technical, so we will omit them, but the results should be believable via the analogy with Morse theory discussed in the preceding section.

The basic workhorse result of this subject is an asymptotic analogue of the similarity principle (Theorem 1.21), in the spirit of Exercise 3.3. To state this, recall that for any closed Reeb orbit $\gamma \colon S^1 \to M$ on a $(2n + 1)$-dimensional contact manifold $(M, \xi = \ker \alpha)$, one can find a *unitary* trivialization of the bundle $\gamma^*\xi \to S^1$, identifying $d\alpha|_\xi$ and $J|_\xi$ with the standard symplectic and complex structures on $\mathbb{R}^{2n} = \mathbb{C}^n$. If $J_0 \colon \mathbb{R}^{2n} \to \mathbb{R}^{2n}$ denotes the standard complex structure, the asymptotic operator $\mathbf{A}_\gamma \colon \Gamma(\gamma^*\xi) \to \Gamma(\gamma^*\xi)$ is then identified with a first-order differential operator

$$\mathbf{A} := -J_0 \frac{d}{dt} - S \colon C^\infty(S^1, \mathbb{R}^{2n}) \to C^\infty(S^1, \mathbb{R}^{2n}), \tag{3.5}$$

where $S \colon S^1 \to \mathrm{End}(\mathbb{R}^{2n})$ is a smooth loop of real $2n$-by-$2n$ matrices, and symmetry of $\mathbf{A}$ with respect to the standard L^2-inner product translates into the condition that $S(t)$ is a symmetric matrix for all t. The following statement and the two that follow it should each be interpreted as *two* closely related statements, one with plus signs and the other with minus signs.

Theorem 3.10 *Suppose* $S \colon Z_\pm \to \mathrm{End}(\mathbb{R}^{2n})$ *is a smooth family of $2n$-by-$2n$ matrices satisfying*

$$S(s,t) \mapsto S(t) \quad \text{uniformly in } t \text{ as } s \to \pm\infty,$$

where $S \colon S^1 \to \mathrm{End}(\mathbb{R}^{2n})$ *is a smooth family of symmetric matrices such that the asymptotic operator $\mathbf{A}$ defined in (3.5) has trivial kernel. Suppose further that* $f \colon Z_\pm \to \mathbb{R}^{2n}$ *is a smooth function that is not identically zero and satisfies*

$$\partial_s f(s,t) + J_0\, \partial_t f(s,t) + S(s,t) f(s,t) = 0, \quad \text{and}$$
$$f(s,\cdot) \to 0 \text{ uniformly as } s \to \pm\infty. \tag{3.6}$$

Then there exists a unique nontrivial eigenfunction $v_\lambda \in C^\infty(S^1, \mathbb{R}^{2n})$ *of $\mathbf{A}$ with*

$$\mathbf{A} v_\lambda = \lambda v_\lambda, \qquad \pm\lambda < 0,$$

and a function $r(s,t) \in \mathbb{R}^{2n}$ *satisfying* $r(s,\cdot) \to 0$ *uniformly as* $s \to \pm\infty$ *such that for sufficiently large $|s|$,*

$$f(s,t) = e^{\lambda s} \left[v_\lambda(t) + r(s,t) \right]. \tag{3.7}$$

Now assume $\gamma \colon S^1 \to M$ is a nondegenerate T-periodic Reeb orbit in $(M, \xi = \ker\alpha)$, parametrized so that $\alpha(\dot\gamma) \equiv T$. Nondegeneracy implies that the asymptotic operator $\mathbf{A}_\gamma$ has trivial kernel. Fixing $J \in \mathcal{J}(\alpha)$, recall that in §2.4, we defined a J-holomorphic half-cylinder $u \colon Z_+ \to \mathbb{R} \times M$ or $u \colon Z_- \to \mathbb{R} \times M$ to be (positively or negatively) asymptotic to γ if, after a possible reparametrization near infinity,

$$u(s,t) = \exp_{(Ts,\gamma(t))} h(s,t) \quad \text{for } |s| \text{ large}, \tag{3.8}$$

where the exponential map is assumed to be translation-invariant and $h(s,t)$ is a vector field along the orbit cylinder with $h(s,\cdot) \to 0$ in $C^\infty(S^1)$ as $s \to \pm\infty$. In particular, as $|s| \to \infty$, $u(s,t)$ becomes C^∞-close to the orbit cylinder $(s,t) \mapsto (Ts, \gamma(t))$, which is an immersion with normal bundle equivalent to $\gamma^*\xi$. After a further reparametrization of $Z_\pm$, we can then arrange for (3.8) to hold for a unique section

$$h(s,t) \in \xi_{\gamma(t)},$$

3.2 Asymptotic Formulas for Cylindrical Ends 51

which we will call the **asymptotic representative** of u. Note that the uniqueness of h depends on our choice of parametrization $\gamma\colon S^1 \to M$ for the Reeb orbit; different choices will change h by a shift in the t-coordinate.

The relation (3.8) is a special case of the following general scenario. We have an almost complex manifold (W, J) with two immersed J-holomorphic curves $v\colon (\Sigma, j) \to (W, J)$ and $u\colon (\Sigma', j') \to (W, J)$, together with a (not necessarily holomorphic) "reparametrization" diffeomorphism $\varphi\colon \Sigma \to \Sigma'$ and a section h of the normal bundle $N_v \to \Sigma$ to v such that

$$u \circ \varphi = \exp_v h, \quad \text{or equivalently} \quad u = \exp_{v \circ \psi} \eta,$$

where we define $\psi := \varphi^{-1}$ and $\eta := h \circ \psi$, the latter being a section of the induced bundle $\psi^* N_v \to \Sigma'$. It turns out (see Proposition B.28 in Appendix B.2.2) that in this situation, one can always view η as a solution of a linear Cauchy–Riemann type equation; hence its local behavior is governed by the similarity principle – or in the asymptotic setting, by Theorem 3.10. In the present context, this idea can be used to prove the following theorem:

Theorem 3.11 *Suppose $u\colon Z_\pm \to \mathbb{R} \times M$ is a J-holomorphic half-cylinder positively/negatively asymptotic to the nondegenerate Reeb orbit $\gamma\colon S^1 \to M$, and let $h_u(s, t) \in \xi_{\gamma(t)}$ denote its asymptotic representative. Then if h_u is not identically zero, there exists a unique nontrivial eigenfunction f_λ of $\mathbf{A}_\gamma$ with*

$$\mathbf{A}_\gamma f_\lambda = \lambda f_\lambda, \quad \pm\lambda < 0,$$

and a section $r(s, t) \in \xi_{\gamma(t)}$ satisfying $r(s, \cdot) \to 0$ uniformly as $s \to \pm\infty$, such that for sufficiently large $|s|$,

$$h_u(s, t) = e^{\lambda s} \left[f_\lambda(t) + r(s, t) \right].$$

In the situation of Theorem 3.11, we will say that $u(s, t)$ approaches the Reeb orbit γ along the **asymptotic eigenfunction** f_λ and with **decay rate** $|\lambda|$. Observe that this theorem can be viewed as describing the asymptotic approach of *two* J-holomorphic half-cylinders to each other, namely $u(s, t)$ and the orbit cylinder $(Ts, \gamma(t))$. A similar result holds for *any* two curves approaching the same orbit, and one can then establish a lower bound on the resulting "relative" decay rate. For our purposes, this result can be expressed most conveniently as follows:

Theorem 3.12 *Suppose $u, v\colon Z_\pm \to \mathbb{R} \times M$ are two J-holomorphic half-cylinders, both positively/negatively asymptotic to the nondegenerate Reeb orbit $\gamma\colon S^1 \to M$, with asymptotic representatives h_u and h_v, asymptotic eigenfunctions f_u, f_v and decay rates $|\lambda_u|$, $|\lambda_v|$, respectively. Then if $h_u - h_v$ is not identically zero, it satisfies*

52   *Asymptotics of Punctured Holomorphic Curves*

$$h_u(s,t) - h_v(s,t) = e^{\lambda s} \left[f_\lambda(t) + r(s,t) \right]$$

for a unique nontrivial eigenfunction f_λ of $\mathbf{A}_\gamma$ with

$$\mathbf{A}_\gamma f_\lambda = \lambda f_\lambda, \qquad \pm\lambda < 0,$$

and a section $r(s,t) \in \xi_{\gamma(t)}$ satisfying $r(s,\cdot) \to 0$ uniformly as $s \to \pm\infty$. Moreover, the following conditions apply:

- *If $f_u = f_v$, then $|\lambda| > |\lambda_u| = |\lambda_v|$.*
- *Otherwise, $|\lambda| = \min\{|\lambda_u|, |\lambda_v|\}$.*

We say in the situation of Theorem 3.12 that u and v approach each other along the **relative asymptotic eigenfunction** f_λ with **relative decay rate** $|\lambda|$.

Observe that if u and v are two asymptotically cylindrical curves with a pair of ends for which $h_u - h_v \equiv 0$ in Theorem 3.12, then standard unique continuation arguments imply that u and v have identical images; i.e., they both cover the same simple curve. In all other cases, the asymptotic formula provides a neighborhood of infinity on which $h_u - h_v$ must be nowhere zero, so u and v have no intersections near infinity. If u and v are asymptotic to different *covers* of the same orbit, then one can argue in the same way by replacing each with suitable covers

$$\tilde{u}(s,t) := u(ks, kt), \qquad \tilde{v}(s,t) := v(\ell s, \ell t),$$

whose asymptotic Reeb orbits match. In this way, one can deduce the following important consequence, which was not previously obvious:

Corollary 3.13   *If $u \colon (\dot{\Sigma}, j) \to (\widehat{W}, J)$ and $v \colon (\dot{\Sigma}', j') \to (\widehat{W}, J)$ are two asymptotically cylindrical J-holomorphic curves with nonidentical images, then they have at most finitely many intersections.*

Similarly,

Corollary 3.14   *If $u \colon (\dot{\Sigma}, j) \to (\widehat{W}, J)$ is an asymptotically cylindrical J-holomorphic curve that is simple, then it is embedded on some neighborhood of the punctures.*

Proof If u has two ends asymptotic to covers of the same orbit, we deduce as in Corollary 3.13 that their images are either identical or disjoint near infinity, and the former is excluded via unique continuation arguments if u is simple. There could still be double points near a single end asymptotic to a multiply covered Reeb orbit; i.e., suppose $Z_\pm \subset \dot{\Sigma}$ is an end on which $u|_{Z_\pm}$ is asymptotic to

$$\gamma(t) = \gamma_0(kt),$$

where $k \geq 2$ is an integer and $\gamma_0 \colon S^1 \to M$ is an *embedded* Reeb orbit. Then, writing $u(s,t) = \exp_{(Ts,\gamma_0(kt))} h(s,t)$ on $Z_{\pm}$ as in Theorem 3.11, the reparametrizations $u_j(s,t) := u(s, t + j/k)$ for $j = 1, \ldots, k-1$ are each also J-holomorphic half-cylinders asymptotic to γ, with asymptotic representatives $h_j(s,t) := h(s, t + j/k)$, and Theorem 3.12 implies that $h_j - h$ is either identically zero or nowhere zero near infinity for $j = 1, \ldots, k-1$. The former is again excluded via unique continuation if u is simple. $\qquad\square$

3.3 Winding of Asymptotic Eigenfunctions

When $\dim M = 3$, the asymptotic eigenfunctions in the above discussion are nowhere vanishing sections of complex line bundles $\gamma^*\xi \to S^1$, so they have well-defined winding numbers relative to any choice of trivialization. This defines the notion of the *asymptotic winding* of a holomorphic curve as it approaches an orbit. It is extremely useful to observe that these winding numbers come with a priori bounds.

Theorem 3.15 ([HWZ95]) *Suppose* $S \colon S^1 \to \text{End}(\mathbb{R}^2)$ *is a smooth loop of symmetric 2-by-2 matrices and* $\mathbf{A} \colon C^\infty(S^1, \mathbb{R}^2) \to C^\infty(S^1, \mathbb{R}^2)$ *denotes the model asymptotic operator*

$$\mathbf{A} = -J_0 \frac{d}{dt} - S,$$

with spectrum $\sigma(\mathbf{A}) \subset \mathbb{R}$. *Then there exists a well-defined integer-valued function*

$$\text{wind} \colon \sigma(\mathbf{A}) \to \mathbb{Z}$$

determined by $\text{wind}(\lambda) := \text{wind}(v_\lambda)$, *where* $v_\lambda \in C^\infty(S^1, \mathbb{R}^2)$ *is any nontrivial eigenfunction with eigenvalue* λ. *Moreover, this function is monotone increasing and attains every value in* $\mathbb{Z}$ *exactly twice (counting multiplicity of eigenvalues).*

Exercise 3.16 Verify Theorem 3.15 for the special case where $S(t)$ is a constant multiple of the identity matrix. (*The general case can be derived from this using perturbation theory for self-adjoint operators; see* [HWZ95, Lemma 3.6] *or* [Wenb, Chapter 3].)

Given a closed Reeb orbit γ in a contact 3-manifold $(M, \xi = \ker \alpha)$, one can now choose a trivialization

$$\tau \colon \gamma^*\xi \to S^1 \times \mathbb{R}^2$$

54 *Asymptotics of Punctured Holomorphic Curves*

and define
$$\text{wind}^\tau \colon \sigma(\mathbf{A}_\gamma) \to \mathbb{Z}$$
by $\text{wind}^\tau(\lambda) := \text{wind}(f)$, where $f \colon S^1 \to \mathbb{R}^2$ is the expression via τ of any nontrivial eigenfunction $f_\lambda \in \Gamma(\gamma^*\xi)$ with $\mathbf{A}_\gamma f_\lambda = \lambda f_\lambda$. It follows immediately from Theorem 3.15 that wind^τ is a monotone surjective function attaining all values exactly twice. Since eigenvalues of $\mathbf{A}_\gamma$ do not accumulate except at $\pm\infty$, we can then define the integers:

$$\alpha_+^\tau(\gamma) := \min \left\{ \text{wind}^\tau(\lambda) \mid \lambda \in \sigma(\mathbf{A}_\gamma) \cap (0, \infty) \right\},$$
$$\alpha_-^\tau(\gamma) := \max \left\{ \text{wind}^\tau(\lambda) \mid \lambda \in \sigma(\mathbf{A}_\gamma) \cap (-\infty, 0) \right\}, \tag{3.9}$$
$$p(\gamma) := \alpha_+^\tau(\gamma) - \alpha_-^\tau(\gamma).$$

As implied by this choice of notation, $\alpha_\pm^\tau(\gamma)$ each depend on the choice of trivialization τ, but $p(\gamma)$ does not. If γ is nondegenerate, hence $0 \notin \sigma(\mathbf{A}_\gamma)$, it follows from Theorem 3.15 that $p(\gamma)$ is either 0 or 1: we shall say accordingly that γ is **even** or **odd**, respectively, and call $p(\gamma)$ the **parity** of γ.

The winding invariants we've just defined have an important relation with another integer associated with nondegenerate Reeb orbits, namely the **Conley–Zehnder index**

$$\mu_{CZ}^\tau(\gamma) \in \mathbb{Z},$$

a Maslov-type index that was originally introduced in the study of Hamiltonian systems (see [CZ83, SZ92]) and can also be defined for nondegenerate Reeb orbits in any dimension. It can be thought of as a measurement of the degree of "twisting" (relative to τ) of the nearby Reeb flow around γ. We refer to [HWZ95, §3] or [Wenb, Chapter 3] for further details on μ_{CZ}^τ; for our purposes in the 3-dimensional case, the following result from [HWZ95, §3] can just as well be taken as a *definition*:

Proposition 3.17 *For any nondegenerate Reeb orbit $\gamma \colon S^1 \to M$ in $(M, \xi = \ker \alpha)$ with a trivialization τ of $\gamma^*\xi$,*

$$\mu_{CZ}^\tau(\gamma) = 2\alpha_-^\tau(\gamma) + p(\gamma) = 2\alpha_+^\tau(\gamma) - p(\gamma).$$

Exercise 3.18 To any closed Reeb orbit of period $T > 0$ parametrized by a loop $\gamma \colon S^1 \to M$ with $\dot{\gamma} \equiv T \cdot R_\alpha(\gamma)$, one can associate a Reeb orbit of period kT for each $k \in \mathbb{N}$, parametrized by

$$\gamma^k \colon S^1 \to M \colon t \mapsto \gamma(kt).$$

We say γ^k is the *k*-**fold cover** of γ, and it is **multiply covered** if $k \geq 2$. We say γ is **simply covered** if it is not the k-fold cover of another Reeb orbit for any $k \geq 2$.

3.4 Local Foliations and the Normal Chern Number 55

(a) Given a Reeb orbit γ, check that the k-fold cover of each eigenfunction of $\mathbf{A}_\gamma$ is an eigenfunction of $\mathbf{A}_{\gamma^k}$. Assuming τ is the pullback under $S^1 \to S^1 : t \mapsto kt$ of a trivialization of $\gamma^* \xi \to S^1$, deduce from Theorem 3.15 that a nontrivial eigenfunction f of $\mathbf{A}_{\gamma^k}$ is a k-fold cover if and only if $\mathrm{wind}^\tau(f)$ is divisible by k.

(b) Under the same assumptions, show that for any nontrivial eigenfunction f of $\mathbf{A}_{\gamma^k}$,

$$\mathrm{cov}(f) := \max\{m \in \mathbb{N} \mid f \text{ is an } m\text{-fold cover}\} = \gcd(k, \mathrm{wind}^\tau(f)).$$

(c) Show that if γ is a Reeb orbit that has even Conley–Zehnder index, then so does every multiple cover γ^k of γ.

3.4 Local Foliations and the Normal Chern Number

We now address a generalization of the question considered in §1.3: if $(\widehat{W}, \omega)$ is a completed symplectic cobordism of dimension 4, what conditions can guarantee that a 2-parameter family of embedded punctured holomorphic curves in $\widehat{W}$ will form a foliation? There are several issues here that do not arise in the closed case: for example, if

$$u : \dot{\Sigma} = \Sigma \setminus (\Gamma^+ \cup \Gamma^-) \to \widehat{W}$$

is embedded, it is not guaranteed in general that all nearby curves $u_\epsilon : \dot{\Sigma} \to \widehat{W}$ are also embedded; e.g., u may have multiple ends asymptotic to the same Reeb orbit, allowing u_ϵ to have double points near that orbit that escape to infinity as $u_\epsilon \to u$. We will address this issue in the next lecture and ignore it for now, as we must first deal with the more basic question of how to count zeroes of sections on the normal bundle $N_u \to \dot{\Sigma}$. Indeed, let us consider as in §1.3 a 1-parameter family of J-holomorphic curves u_σ near $u_0 := u$, presented as $\exp_u \eta_\sigma$ for sections $\eta_\sigma \in \Gamma(N_u)$. One can then show that

$$\eta := \left. \frac{\partial}{\partial \sigma} u_\sigma \right|_{\sigma=0} \in \Gamma(N_u) \tag{3.10}$$

satisfies a linear Cauchy–Riemann type equation. We would like to know when such sections are guaranteed to be nowhere zero. Write the positive and negative contact boundary components of the cobordism (W, ω) as

$$\partial(W, \omega) = (-M_-, \xi_-) \sqcup (M_+, \xi_+).$$

Since u is always transverse to the contact bundles $\xi_\pm$ near infinity, one can identify N_u with $u^* \xi_\pm$ on the cylindrical ends. By the similarity principle, zeroes of η are isolated and positive, but the total algebraic count of them is

56 *Asymptotics of Punctured Holomorphic Curves*

not a homotopy invariant since they may escape to infinity under homotopies; in fact, there could in theory be infinitely many. It turns out, however, that on any cylindrical end $Z_\pm \subset \dot{\Sigma}$ near a puncture $z \in \Gamma^\pm$, where u is asymptotic to an orbit γ_z, the relevant linear Cauchy–Riemann type equation has the same form as in Theorem 3.10, with $\mathbf{A}_{\gamma_z}$ as the relevant asymptotic operator. The theorem thus implies that η is nowhere zero near each puncture z, and it has a well-defined **asymptotic winding** relative to any choice of trivialiation τ of $\gamma_z^* \xi_\pm$,

$$\mathrm{wind}^\tau(\eta; z) \in \mathbb{Z},$$

defined simply as $\mathrm{wind}^\tau(v_\lambda)$, where $v_\lambda \in \Gamma(\gamma_z^* \xi_\pm)$ is the asymptotic eigenfunction appearing in (3.7). This implies that $\eta^{-1}(0) \subset \dot{\Sigma}$ is finite, so we can define the algebraic count of zeroes

$$Z(\eta) := \sum_{z \in \eta^{-1}(0)}, \mathrm{ord}(\eta; z) \in \mathbb{Z}, \tag{3.11}$$

where $\mathrm{ord}(\eta; z)$ denotes the order of each zero, and the similarity principle guarantees that $Z(\eta) \geq 0$, with equality if and only if η is nowhere zero. This number is still not homotopy invariant, because zeroes can still escape to infinity under homotopies. However, the crucial observation is that we can *keep track* of this phenomenon via the asymptotic winding numbers: by Theorem 3.15, $\mathrm{wind}^\tau(\eta; z)$ satisfies the a priori bounds

$$\begin{aligned}
\mathrm{wind}^\tau(\eta; z) &\leq \alpha_-^\tau(\gamma_z), &&\text{if } z \in \Gamma^+, \\
\mathrm{wind}^\tau(\eta; z) &\geq \alpha_+^\tau(\gamma_z), &&\text{if } z \in \Gamma^-.
\end{aligned} \tag{3.12}$$

This motivates the definition of the **asymptotic defect** of η as the integer

$$Z_\infty(\eta) := \sum_{z \in \Gamma^+} \left[\alpha_-^\tau(\gamma_z) - \mathrm{wind}^\tau(\eta; z) \right] + \sum_{z \in \Gamma^-} \left[\mathrm{wind}^\tau(\eta; z) - \alpha_+^\tau(\gamma_z) \right], \tag{3.13}$$

where the trivializations τ of $\gamma_z^* \xi_\pm$ can be chosen arbitrarily, since each difference $\alpha_\mp^\tau(\gamma_z) - \mathrm{wind}^\tau(\eta; z)$ does not depend on this choice. By construction, any $\eta \in \Gamma(N_u)$ satisfying a Cauchy–Riemann type equation as described above now has both $Z(\eta) \geq 0$ and $Z_\infty(\eta) \geq 0$, and their sum turns out to give the closest thing possible to a homotopy-invariant count of zeroes:

Proposition 3.19 *For any section $\eta \in \Gamma(N_u)$ with only finitely many zeroes, the sum $Z(\eta) + Z_\infty(\eta)$ depends only on the bundle N_u and the asymptotic operators $\mathbf{A}_z$ for $z \in \Gamma$, not on η. In particular, this gives an upper bound on the algebraic count of zeroes of any section η appearing in (3.10).*

This result motivates the interpretation of $Z_\infty(\eta)$ as a count of *virtual* or "hidden zeroes at infinity." We will prove Proposition 3.19 by defining another

3.4 Local Foliations and the Normal Chern Number 57

quantity that is manifestly homotopy invariant and happens to equal $Z(\eta)$ + $Z_\infty(\eta)$: this will be a generalization of the *normal Chern number*, which we defined for closed holomorphic curves in §2.1.

We must first define the notion of a *relative* first Chern number for complex vector bundles over punctured surfaces. Suppose first that $E \to \dot{\Sigma}$ is a complex line bundle, and τ denotes a choice of trivializations for E over the cylindrical ends of $\dot{\Sigma}$, i.e., over small neighborhoods of each puncture. Such trivializations always exist since complex vector bundles over S^1 are always trivial. In fact, $E \to \dot{\Sigma}$ is globally trivializable if the set of punctures is nonempty, because $\dot{\Sigma}$ is then retractable to its 1-skeleton – nonetheless, a given set of trivializations τ over the ends may or may not be globally extendable over the rest of $\dot{\Sigma}$. An obstruction to such extensions is given by the **relative first Chern number** of E with respect to τ: we define it as an algebraic count of zeroes,

$$c_1^\tau(E) := Z(\eta) \in \mathbb{Z},$$

where $Z(\eta)$ is defined as in (3.11) for a section $\eta \in \Gamma(E)$ with finitely many zeroes, and we assume that η is *constant and nonzero near infinity* with respect to τ. It follows by standard arguments as in [Mil97] that $c_1^\tau(E)$ does not depend on the choice η: the point is that any two such choices are homotopic through sections that are nonzero near infinity, so zeroes stay within a compact subset under the homotopy. Observe that in the special case where $\dot{\Sigma} = \Sigma$ is a closed surface without punctures, there is no choice of asymptotic trivialization τ to be made and the above definition matches the usual first Chern number $c_1(E)$. When there are punctures, $c_1^\tau(E)$ depends on the choice τ.

For a higher-rank complex vector bundle $E \to \dot{\Sigma}$ with a trivialization τ near infinity, $c_1^\tau(E)$ can be defined by assuming the following two axioms:

(1) $c_1^{\tau_1 \oplus \tau_2}(E_1 \oplus E_2) = c_1^{\tau_1}(E_1) + c_1^{\tau_2}(E_2)$.
(2) $c_1^\tau(E) = c_1^{\tau'}(E')$ whenever E and E' admit a complex bundle isomorphism identifying τ with τ'.

The following exercise shows that this is a reasonable definition.

Exercise 3.20 Show that for any complex vector bundle E over a punctured Riemann surface $\dot{\Sigma}$ of rank n with an asymptotic trivialization τ, there exist complex line bundles $E_1, \ldots, E_n \to \dot{\Sigma}$ with asymptotic trivializations $\tau_1, \ldots, \tau_n$ such that

$$(E, \tau) \cong (E_1 \oplus \cdots \oplus E_n, \tau_1 \oplus \cdots \oplus \tau_n),$$

and if $E_1', \ldots, E_n'$ and $\tau_1', \ldots, \tau_n'$ are another n-tuple of line bundles and asymptotic trivializations with this property, then

Asymptotics of Punctured Holomorphic Curves

$$c_1^{\tau_1}(E_1) + \cdots + c_1^{\tau_2}(E_2) = c_1^{\tau_1'}(E_1') + \cdots + c_1^{\tau_n'}(E_n').$$

From now on, let τ denote a fixed arbitrary choice of trivializations of the bundles $\gamma^*\xi_\pm$ for all Reeb orbits γ; several things in the calculations below will depend on this choice, but the most important expressions typically will not. Since the normal bundle N_u matches $\xi_\pm$ near infinity, τ determines an asymptotic trivialization of N_u, allowing us to define the relative first Chern number $c_1^\tau(N_u)$. More generally, if $u \colon \dot{\Sigma} \to \widehat{W}$ is *any* asymptotically cylindrical map, not necessarily immersed, then it is still immersed and transverse to $\xi_\pm$ near infinity, so τ also determines an asymptotic trivialization of the rank 2 complex vector bundle $(u^*T\widehat{W}, J) \to \dot{\Sigma}$ by observing that the first factor in the splitting

$$T(\mathbb{R} \times M_\pm) = (\mathbb{R} \oplus \mathbb{R}R_{\alpha_\pm}) \oplus \xi_\pm$$

carries a canonical complex trivialization. We shall denote the resulting relative first Chern number for $u^*T\widehat{W}$ by $c_1^\tau(u^*T\widehat{W})$.

Exercise 3.21 Show that if $u \colon (\dot{\Sigma}, j) \to (\widehat{W}, J)$ is an asymptotically cylindrical and immersed J-holomorphic curve, with complex normal bundle $N_u \to \dot{\Sigma}$, then

$$c_1^\tau(u^*T\widehat{W}) = \chi(\dot{\Sigma}) + c_1^\tau(N_u). \tag{3.14}$$

Definition 3.22 For any asymptotically cylindrical J-holomorphic curve $u \colon (\dot{\Sigma}, j) \to (\widehat{W}, J)$ asymptotic to Reeb orbits γ_z in $M_\pm$ at its punctures $z \in \Gamma^\pm$, we define the **normal Chern number** of u to be the integer

$$c_N(u) := c_1^\tau(u^*T\widehat{W}) - \chi(\dot{\Sigma}) + \sum_{z \in \Gamma^+} \alpha_-^\tau(\gamma_z) - \sum_{z \in \Gamma^-} \alpha_+^\tau(\gamma_z).$$

Exercise 3.23 Show that the definition of $c_N(u)$ above is independent of the choice of trivializations τ.

The normal Chern number $c_N(u)$ clearly depends only on the homotopy class of u as an asymptotically cylindrical map, together with the properties of its asymptotic Reeb orbits. When u is immersed, we can rewrite it via (3.14) as

$$c_N(u) = c_1^\tau(N_u) + \sum_{z \in \Gamma^+} \alpha_-^\tau(\gamma_z) - \sum_{z \in \Gamma^-} \alpha_+^\tau(\gamma_z). \tag{3.15}$$

Proposition 3.19 then follows immediately from the following theorem:

Theorem 3.24 *Suppose* $u \colon (\dot{\Sigma}, j) \to (\widehat{W}, J)$ *is an immersed asymptotically cylindrical J-holomorphic curve, and* $\eta \in \Gamma(N_u)$ *is a smooth section of its normal bundle with at most finitely many zeroes. Then*

$$Z(\eta) + Z_\infty(\eta) = c_N(u).$$

3.4 Local Foliations and the Normal Chern Number 59

In the situation of interest, we already know that both $Z(\eta)$ and $Z_\infty(\eta)$ are nonnegative, so this yields the following corollary:

Corollary 3.25 *If $u \colon (\dot{\Sigma}, j) \to (\widehat{W}, J)$ is an immersed asymptotically cylindrical J-holomorphic curve and $\eta \in \Gamma(N_u)$ is a section of its normal bundle describing nearby J-holomorphic curves as in* (3.10), *then*

$$Z(\eta) \le c_N(u);$$

in particular, if $c_N(u) = 0$, then every such section is zero free.

Proof of Theorem 3.24 Let τ_0 denote the unique choice of asymptotic trivialization of N_u such that

$$\operatorname{wind}^{\tau_0}(\eta; z) = 0 \quad \text{for all } z \in \Gamma.$$

Note that if u has multiple ends approaching the same orbit γ in $M_\pm$, this choice may require nonisomorphic trivializations of $\gamma^* \xi_\pm$ for different ends, but this will pose no difficulty in the following. For this choice, we have

$$Z(\eta) = c_1^{\tau_0}(N_u);$$

thus, using (3.15) and the definition (3.13) of $Z_\infty(\eta)$,

$$Z(\eta) + Z_\infty(\eta) = c_1^{\tau_0}(N_u) + \sum_{z \in \Gamma^+} \alpha_-^{\tau_0}(\gamma_z) - \sum_{z \in \Gamma^-} \alpha_+^{\tau_0}(\gamma_z)$$

$$= c_N(u). \qquad \square$$

Corollary 3.25 tells us that in order to find 2-dimensional families of embedded J-holomorphic curves that locally form foliations, one should restrict attention to curves satisfying $c_N(u) = 0$. To see what kinds of curves satisfy this condition, recall that a general J-holomorphic curve $u \colon \dot{\Sigma} \to \widehat{W}$ in a $2n$-dimensional cobordism $\widehat{W}$, with positive/negative punctures $z \in \Gamma := \Gamma^+ \cup \Gamma^-$ asymptotic to nondegenerate Reeb orbits γ_z, is defined to have **index**

$$\operatorname{ind}(u) = (n-3)\chi(\dot{\Sigma}) + 2c_1^\tau(u^* T\widehat{W}) + \sum_{z \in \Gamma^+} \mu_{\mathrm{CZ}}^\tau(\gamma_z) - \sum_{z \in \Gamma^-} \mu_{\mathrm{CZ}}^\tau(\gamma_z)$$

(see Appendix A.2). As usual, all dependence on the trivialization τ in terms on the right-hand side cancels out in the sum. This index is the *virtual dimension* of the moduli space of all curves homotopic to u, and, for generic J, the open subset of simple curves in this space is a smooth manifold of this dimension. Let us restrict to the case $\dim \widehat{W} = 4$, so $n = 2$, and let g denote the genus of Σ; hence

60 · Asymptotics of Punctured Holomorphic Curves

$$\mathrm{ind}(u) = -\chi(\dot{\Sigma}) + 2c_1^\tau(u^*T\widehat{W}) + \sum_{z\in\Gamma^+}\mu_{\mathrm{CZ}}^\tau(\gamma_z) - \sum_{z\in\Gamma^-}\mu_{\mathrm{CZ}}^\tau(\gamma_z), \qquad (3.16)$$

and

$$\chi(\dot{\Sigma}) = 2 - 2g - \#\Gamma. \qquad (3.17)$$

There is also a natural partition of Γ into the *even* and *odd* punctures

$$\Gamma = \Gamma_{\mathrm{even}} \cup \Gamma_{\mathrm{odd}},$$

defined via the parity of the corresponding orbit as defined in §3.3, or, equivalently, the parity of the Conley–Zehnder index.[1] Now combining (3.16), (3.17), Definition 3.22 and the Conley–Zehnder/winding relations of Proposition 3.17, we have

$$
\begin{aligned}
2c_N(u) &= 2c_1^\tau(u^*T\widehat{W}) - 2\chi(\dot{\Sigma}) + \sum_{z\in\Gamma^+}2\alpha_-^\tau(\gamma_z) - \sum_{z\in\Gamma^-}2\alpha_+^\tau(\gamma_z) \\
&= 2c_1^\tau(u^*T\widehat{W}) - \chi(\dot{\Sigma}) - (2 - 2g - \#\Gamma) + \sum_{z\in\Gamma^+}\left[\mu_{\mathrm{CZ}}^\tau(\gamma_z) - p(\gamma_z)\right] \\
&\quad - \sum_{z\in\Gamma^-}\left[\mu_{\mathrm{CZ}}^\tau(\gamma_z) + p(\gamma_z)\right] \\
&= \mathrm{ind}(u) - 2 + 2g + \#\Gamma - \#\Gamma_{\mathrm{odd}} \\
&= \mathrm{ind}(u) - 2 + 2g + \#\Gamma_{\mathrm{even}}.
\end{aligned}
\qquad (3.18)
$$

Since we are interested in 2-dimensional families of curves, assume $\mathrm{ind}(u) = 2$. Then the right-hand side of (3.18) is nonnegative and vanishes if and only if $g = \#\Gamma_{\mathrm{even}} = 0$; i.e., $\dot{\Sigma}$ is a punctured *sphere* and all asymptotic orbits have *odd* Conley–Zehnder index. This leads to the following result. We state it for now with an extra assumption (condition (4) in Theorem 3.26) in order to avoid the possibility of extra intersections emerging from infinity – this can be relaxed using the technology introduced in the next lecture, but the weaker result will also suffice for our application in Lecture 5.

Theorem 3.26 *Suppose* $u\colon (\dot{\Sigma}, j) \to (\widehat{W}, J)$ *is an embedded asymptotically cylindrical J-holomorphic curve such that*

(1) $\mathrm{ind}(u) = 2$.
(2) $\dot{\Sigma}$ *has genus* 0.

---

[1] Note that while the Conley–Zehnder index $\mu_{\mathrm{CZ}}^\tau(\gamma) \in \mathbb{Z}$ generally depends on a choice of trivialization τ of the contact bundle along γ, different choices of trivialization change the index by multiples of 2; thus the odd/even parity is independent of this choice.

3.4 Local Foliations and the Normal Chern Number

(3) *All asymptotic orbits of u have odd Conley–Zehnder index.*

(4) *All the punctures are asymptotic to distinct Reeb orbits, all of them simply covered.*

Then some neighborhood of u in the moduli space $\mathcal{M}_0(\widehat{W}, J)$ is a smooth 2-dimensional manifold consisting of pairwise disjoint embedded curves that foliate a neighborhood of $u(\dot{\Sigma})$ in $\widehat{W}$. □

Lecture 4

Intersection Theory for Punctured Holomorphic Curves

We are now ready to explain the intersection theory introduced by Siefring [Sie11] for asymptotically cylindrical holomorphic curves in 4-dimensional completed symplectic cobordisms. The theory follows a pattern that we saw in our discussion of the normal Chern number in §3.4: the obvious geometrically meaningful quantities such as $u \cdot v$ (counting intersections between u and v) and $\delta(u)$ (counting double points and critical points of u) can be defined, and are nonnegative, but they are not homotopy invariant since intersections may sometimes escape to infinity. In each case, however, one can add a nonnegative count of "hidden intersections at infinity," defined in terms of asymptotic winding numbers, so that the sum is homotopy invariant.

4.1 Statement of the Main Results

Throughout this lecture, we assume (W, ω) is a 4-*dimensional symplectic cobordism* with $\partial(W, \omega) = (-M_-, \xi_- = \ker \alpha_-) \sqcup (M_+, \xi_+ = \ker \alpha_+)$, $(\widehat{W}, \omega)$ is its completion and $J \in \mathcal{J}(\omega, \alpha_+, \alpha_-)$. For two asymptotically cylindrical maps $u \colon \dot{\Sigma} \to \widehat{W}$ and $v \colon \dot{\Sigma}' \to \widehat{W}$ with at most finitely many intersections, we define the algebraic intersection number

$$u \cdot v := \sum_{u(z)=v(\zeta)} \iota(u, z \, ; \, v, \zeta) \in \mathbb{Z},$$

and, similarly, if u has at most finitely many double points and critical points, then it has a well-defined *singularity index*

$$\delta(u) := \frac{1}{2} \sum_{u(z)=u(\zeta), \, z \neq \zeta} \iota(u, z \, ; \, u, \zeta) + \sum_{du(z)=0} \delta(u, z) \in \mathbb{Z},$$

4.1 Statement of the Main Results 63

i.e., the sum of the local intersection indices for all double points with the local singularity index at each critical point (cf. Lemma 2.6). If u and v are both asymptotically cylindrical J-holomorphic curves, then as we saw in Corollaries 3.13 and 3.14, $u \cdot v$ is well defined if u and v have nonidentical images, and $\delta(u)$ is also well defined if u is simple. Moreover, the usual results on positivity of intersections (Appendix B) then imply

$$u \cdot v \geq 0,$$

with equality if and only if u and v are disjoint, and

$$\delta(u) \geq 0,$$

with equality if and only if u is embedded. So far, all of this is the same as in the closed case, but the crucial difference here is that neither $u \cdot v$ nor $\delta(u)$ is invariant under homotopies, which makes them harder to control in general. For example, there is no reasonable definition of "$u \cdot u$," since trying to count intersections of u with a small perturbation of itself (as one does in the closed case) may give a number that depends on the perturbation. The situation is saved by the following results from [Sie11]:

Theorem 4.1 *For any two asymptotically cylindrical maps* $u \colon \dot{\Sigma} \to \widehat{W}$ *and* $v \colon \dot{\Sigma}' \to \widehat{W}$ *with nondegenerate[1] asymptotic orbits, there exists a pairing*

$$u * v \in \mathbb{Z}$$

with the following properties:

(1) $u * v$ *depends only on the homotopy classes of* u *and* v *as asymptotically cylindrical maps.*
(2) *If* $u \colon (\dot{\Sigma}, j) \to (\widehat{W}, J)$ *and* $v \colon (\dot{\Sigma}, j') \to (\widehat{W}, J)$ *are* J-holomorphic curves *with nonidentical images, then*

$$u * v = u \cdot v + \iota_\infty(u, v),$$

where $\iota_\infty(u, v)$ *is a nonnegative integer interpreted as the count of "hidden intersections at infinity." Moreover, there exists a perturbation* $J_\epsilon \in \mathcal{J}(\omega, \alpha_+, \alpha_-)$ *which is* C^∞-close to J, *and a pair of asymptotically*

---

[1] Theorems 4.1 and 4.4 both also hold under the more general assumption that all asymptotic orbits belong to Morse–Bott families, as long as one imposes the restriction that asymptotic orbits of curves are not allowed to change under homotopies. (This assumption is vacuous in the nondegenerate case since nondegenerate orbits are isolated.) One can also generalize the theory further to allow homotopies with moving asymptotic orbits, in which case additional nonnegative counts of "hidden" intersections must be introduced (see [Wen10a, §4.1] and [SW]).

64 *Intersection Theory for Punctured Holomorphic Curves*

cylindrical J_ϵ-holomorphic curves $u_\epsilon \colon (\dot{\Sigma}, j_\epsilon) \to (\widehat{W}, J_\epsilon)$ and $v_\epsilon \colon (\dot{\Sigma}, j'_\epsilon) \to$
$(\widehat{W}, J_\epsilon)$ close to u and v in their respective moduli spaces, such that

$$u_\epsilon \cdot v_\epsilon = u * v.$$

The last statement in the above theorem, involving the perturbations u_ϵ and v_ϵ, helps us interpret $u * v$ as the count of intersections between *generic* curves homotopic to u and v. That particular detail is not proved in [Sie11], nor anywhere else in the literature – it has the status of a "folk theorem," meaning that at least a few experts would be able to prove it as an exercise, but have not written down the details in any public forum. The proof involves Fredholm theory on exponentially weighted Sobolev spaces, as explained, e.g., in [HWZ99, Wen10a], and we will not prove it here either, but have included the statement mainly for the sake of intuition. It is not needed for any of the most important applications of Theorem 4.1 such as the following:

Corollary 4.2 *If u and v are J-holomorphic curves satisfying $u * v = 0$, then any two J-holomorphic curves that have nonidentical images and are homotopic to u and v respectively are disjoint.*

In order to write down the punctured version of the adjunction formula, we must introduce a little bit more notation. Suppose $\gamma \colon S^1 \to M$ is a Reeb orbit in a contact 3-manifold $(M, \xi = \ker \alpha)$, and $k \in \mathbb{N}$. This gives rise to the k-fold covered Reeb orbit

$$\gamma^k \colon S^1 \to M \colon t \mapsto \gamma(kt),$$

and we define the **covering multiplicity** $\mathrm{cov}(\gamma)$ of a general Reeb orbit γ as the largest $k \in \mathbb{N}$, such that $\gamma = \gamma_0^k$ for some other Reeb orbit γ_0. Similarly, if $f \in \Gamma(\gamma^* \xi)$ is an eigenfunction of $\mathbf{A}_\gamma$ with eigenvalue $\lambda \in \mathbb{R}$, then the k-fold cover

$$f^k \in \Gamma((\gamma^k)^* \xi), \qquad f^k(t) := f(kt)$$

is an eigenfunction of $\mathbf{A}_{\gamma^k}$ with eigenvalue $k\lambda$, and for any Reeb orbit γ and nontrivial eigenfunction f of $\mathbf{A}_\gamma$, we define $\mathrm{cov}(f) \in \mathbb{N}$ to be the largest integer k, such that f is a k-fold cover of an eigenfunction for a Reeb orbit covered by γ. Observe that, in general, $1 \leq \mathrm{cov}(f) \leq \mathrm{cov}(\gamma)$, and $\mathrm{cov}(f)$ always divides $\mathrm{cov}(\gamma)$. Note also that any trivialization τ of $\gamma^* \xi$ naturally determines a trivialization of $(\gamma^k)^* \xi$, which we shall denote by τ^k.

Remark 4.3 Exercise 3.18 implies that if $\gamma \colon S^1 \to M$ is a simply covered (i.e., embedded) Reeb orbit in a contact 3-manifold $(M, \xi = \ker \alpha)$ and τ is a trivialization of $\gamma^* \xi$, then for any $k \in \mathbb{N}$ and a nontrivial eigenfunction f of $\mathbf{A}_{\gamma^k}$ with $\mathrm{wind}^{\tau^k}(f) > 0$, $\mathrm{cov}(f)$ depends only on k and $\mathrm{wind}^{\tau^k}(f)$; in fact:

4.1 Statement of the Main Results

$$\mathrm{cov}(f) = \gcd(k, \mathrm{wind}^{\tau^k}(f)).$$

We now associate to any Reeb orbit γ in a contact 3-manifold $(M, \xi = \ker \alpha)$ the **spectral covering numbers**

$$\bar{\sigma}_\pm(\gamma) := \mathrm{cov}(f_\pm) \in \mathbb{N},$$

where $f_\pm \in \Gamma(\gamma^* \xi)$ is any choice of eigenfunction of $\mathbf{A}_\gamma$ with $\mathrm{wind}^\tau(f_\pm) = \alpha_\pm^\tau(\gamma)$. Remark 4.3 implies that $\bar{\sigma}_\pm(\gamma)$ does not depend on this choice. Finally, if $u : \dot{\Sigma} \to \widehat{W}$ is an asymptotically cylindrical map with punctures $z \in \Gamma^\pm$ asymptotic to orbits γ_z in $M_\pm$, we define the **total spectral covering number** of u by

$$\bar{\sigma}(u) := \sum_{z \in \Gamma^+} \bar{\sigma}_-(\gamma_z) + \sum_{z \in \Gamma^-} \bar{\sigma}_+(\gamma_z).$$

Observe that $\bar{\sigma}(u)$ really does not depend on the map u, but only on its sets of positive and negative asymptotic orbits. It is a positive integer in general, and we have

$$\bar{\sigma}(u) - \#\Gamma \geq 0,$$

with equality if and only if all of the so-called "extremal" eigenfunctions at the asymptotic orbits of u are simply covered. This is true in particular whenever all asymptotic orbits of u are simply covered.

The next statement is the punctured generalization of the adjunction formula (Theorem 2.8): it relates $u * u$ to $\delta(u)$, the spectral covering number $\bar{\sigma}(u)$, and our generalization of the normal Chern number $c_N(u)$ from §3.4 (see Definition 3.22).

Theorem 4.4 *If $u : (\dot{\Sigma}, j) \to (\widehat{W}, J)$ is an asymptotically cylindrical and simple J-holomorphic curve with punctures $\Gamma \subset \Sigma$, then there exists an integer*

$$\delta_\infty(u) \geq 0,$$

interpreted as the count of "hidden double points at infinity," such that

$$u * u = 2\left[\delta(u) + \delta_\infty(u)\right] + c_N(u) + \left[\bar{\sigma}(u) - \#\Gamma\right]. \tag{4.1}$$

In particular, $\delta(u) + \delta_\infty(u)$ depends only on the homotopy class of u as an asymptotically cylindrical map. Moreover, there exists a perturbation $J_\epsilon \in \mathcal{J}(\omega, \alpha_+, \alpha_-)$ that is C^∞-close to J, and a J_ϵ-holomorphic curve $u_\epsilon : (\dot{\Sigma}, j_\epsilon) \to (\widehat{W}, J_\epsilon)$ close to u in the moduli space, such that $\delta_\infty(u_\epsilon) = 0$.

Corollary 4.5 *If $u \in \mathcal{M}_g(\widehat{W}, J)$ is simple and satisfies $\delta(u) = \delta_\infty(u) = 0$, then every simple curve in the same connected component of $\mathcal{M}_g(\widehat{W}, J)$ is embedded.*

66 *Intersection Theory for Punctured Holomorphic Curves*

Remark 4.6 It is important to notice the lack of the words "and only if" in Corollary 4.5: an embedded curve u always has $\delta(u) = 0$ but may in general have $\delta_\infty(u) > 0$, in which case it could be homotopic to a simple curve with critical or double points.

The remainder of this lecture will be concerned with the definitions of $u * v$, $\iota_\infty(u, v)$ and $\delta_\infty(u)$ and the proofs of Theorems 4.1 and 4.4.

Remark 4.7 The reader should be aware of a few notational differences between this book and the original source [Sie11]. One relatively harmless difference is in the appearance of the adunction formula (Equation (4.1) vs. [Sie11, Equation (2-5)]), as Siefring does not define or mention the normal Chern number but writes an expression that is equivalent due to (3.18). A more serious difference of conventions appears in the formulas we will use to define $u * v$ and $\delta(u)$ later; e.g., (4.4) and (4.12) contain "$\pm$" and "$\mp$" symbols that do not appear in the equivalent formulas in [Sie11]. The reason is that alternate versions of these numbers need to be defined for asymptotic orbits that appear at positive or negative ends; Siefring handles this issue with a notational shortcut, formally viewing Reeb orbits that occur at negative ends as orbits with *negative covering multiplicity*. In these lectures, covering multiplicities are always positive.

4.2 Relative Intersection Numbers and the $*$-Pairing

For the remainder of this lecture, fix a choice of trivializations of the bundles $\gamma^* \xi_\pm \to S^1$ for every *simply covered* Reeb orbit $\gamma \colon S^1 \to M_\pm$. Wherever a trivialization along a multiply covered orbit γ^k is needed, we will use the one induced on $(\gamma^k)^* \xi_\pm \to S^1$ by our chosen trivialization of $\gamma^* \xi_\pm$, and denote this choice as usual by τ.

For two asymptotically cylindrical maps $u \colon \dot{\Sigma} \to \widehat{W}$ and $v \colon \dot{\Sigma}' \to \widehat{W}$, we define the **relative intersection number**

$$u \bullet_\tau v := u \cdot v^\tau \in \mathbb{Z},$$

where $v^\tau \colon \dot{\Sigma}' \to \widehat{W}$ denotes any C^0-small perturbation of v, such that u and v^τ have at most finitely many intersections and v^τ is "pushed off" near $\pm\infty$ in directions determined by τ; i.e., if v approaches the orbit $\gamma \colon S^1 \to M_\pm$ asymptotically at a puncture z, then v^τ at the same puncture approaches a loop of the form $\exp_{\gamma(t)} \epsilon\eta(t)$, where $\epsilon > 0$ is small and $\eta \in \Gamma(\gamma^*\xi_\pm)$ satisfies $\mathrm{wind}^\tau(\eta) = 0$. Since v^τ asymptotically approaches loops that may (without loss of generality) be assumed disjoint from the asymptotic orbits of u, it follows

4.2 Relative Intersection Numbers and the ∗-Pairing 67

from Exercise 2.1 that this definition is independent of the choice of perturbation, and it only depends on the homotopy classes of u and v (as asymptotically cylindrical maps) plus the trivializations τ. The dependence on τ indicates that $u \bullet_\tau v$ is not a very meaningful number on its own, so it will not be an object of primary study for us, but like the relative first Chern numbers in §3.4, it will provide a useful tool for organizing information.

Exercise 4.8 Show that $u \bullet_\tau v = v \bullet_\tau u$.

Suppose $u \colon \dot{\Sigma} \to \widehat{W}$ and $v \colon \dot{\Sigma}' \to \widehat{W}$ are asymptotically cylindrical and have finitely many intersections, so $u \cdot v$ is well defined. Then $u \bullet_\tau v$ can be computed with the perturbation v^τ assumed to be nontrivial only in some neighborhood of infinity where u and v are disjoint, so that $u \cdot v^\tau$ counts the intersections of u with v, plus some additional intersections that appear in a neighborhood of infinity when v is perturbed to v^τ. We shall denote this count of additional intersections near infinity by $\iota_\infty^\tau(u, v) \in \mathbb{Z}$, so we can write

$$u \bullet_\tau v = u \cdot v + \iota_\infty^\tau(u, v)$$

whenever $u \cdot v$ is well defined.

The number $\iota_\infty^\tau(u, v)$ also depends on τ and is thus not meaningful on its own, but it is useful to observe that it can be computed in terms of relative asymptotic winding numbers – this observation will lead us to the natural definitions of the much more meaningful quantities $u \ast v$ and $\iota_\infty(u, v)$, which do not depend on τ. To see this, denote the punctures of u and v by $\Gamma_u = \Gamma_u^+ \cup \Gamma_u^-$ and $\Gamma_v = \Gamma_v^+ \cup \Gamma_v^-$, respectively, and for any $z \in \Gamma_u$ or Γ_v, denote the corresponding asymptotic orbit of u or v by $\gamma_z^{k_z}$, where we assume γ_z is a *simply covered* orbit and $k_z \in \mathbb{N}$ is the covering multiplicity. A contribution to $\iota_\infty^\tau(u, v)$ may come from any pair of punctures $(z, \zeta) \in \Gamma_u^\pm \times \Gamma_v^\pm$, so we shall denote this contribution by $\iota_\infty^\tau(u, z; v, \zeta)$ and write

$$\iota_\infty^\tau(u, v) = \sum_{(z,\zeta) \in \Gamma_u^\pm \times \Gamma_v^\pm} \iota_\infty^\tau(u, z; v, \zeta). \tag{4.2}$$

Remark 4.9 In (4.2) and several other expressions in this lecture, the summation should be understood as a sum of two summations, one with $\pm = +$ and the other with $\pm = -$.

If $\gamma_z \neq \gamma_\zeta$, then u and v^τ have no intersections in neighborhoods of these particular punctures, implying

$$\iota_\infty^\tau(u, z; v, \zeta) = 0 \quad \text{if } \gamma_z \neq \gamma_\zeta.$$

68 *Intersection Theory for Punctured Holomorphic Curves*

Now assume $\gamma := \gamma_z = \gamma_\zeta$, and let $T > 0$ denote the period of γ. We shall parametrize punctured neighborhoods of z and ζ by half-cylinders $Z_\pm$ and consider the resulting maps

$$u(s,t), v(s,t) \in \mathbb{R} \times M_\pm,$$

defined for $|s|$ sufficiently large and asymptotic to γ^{k_z} and γ^{k_ζ}, respectively. We first consider the special case where both asymptotic orbits have the same covering multiplicity, so let

$$k := k_z = k_\zeta.$$

The asymptotic approach to γ^k means we can write

$$u(s,t) = \exp_{(kTs,\gamma(kt))} h_u(s,t), \qquad v(s,t) = \exp_{(kTs,\gamma(kt))} h_v(s,t)$$

for sections h_u and h_v of $\xi_\pm$ along the orbit cylinder such that both decay uniformly to 0 as $s \to \pm\infty$. The assumption that u and v have no intersections near infinity implies, moreover, that for some $s_0 > 0$, each of the sections

$$(s,t) \mapsto h_u(s, t + j/k) - h_v(s,t), \qquad j = 0, \ldots, k-1$$

has no zeroes in the region $|s| \geq s_0$. The perturbation v^τ can now be defined as

$$v^\tau(s,t) = \exp_{(kTs,\gamma(kt))} \left[h_v(s,t) + \epsilon\eta(s,t) \right],$$

where $\epsilon > 0$ is small and $\eta(s,t) \in \xi_\pm$ can be assumed to vanish for $|s| \leq s_0$ and to satisfy $\eta(s,t) \to \eta_\infty(kt)$ as $s \to \pm\infty$, with $\eta_\infty \in \Gamma(\gamma^*\xi_\pm)$ a nowhere vanishing section satisfying $\mathrm{wind}^\tau(\eta_\infty) = 0$. Intersections of v^τ with u in the region $|s| \geq s_0$ are now in one-to-one correspondence with solutions of the equation

$$F_j(s,t) := h_u(s, t + j/k) - h_v(s,t) - \epsilon\eta(s,t) = 0,$$

for arbitrary values of $j \in \{0, \ldots, k-1\}$. Notice that F_j admits a continuous extension to $s = \pm\infty$ with $F_j(\pm\infty, t) = -\epsilon\eta_\infty(kt)$. Since $\mathrm{wind}^\tau(\eta_\infty) = 0$ and $\epsilon > 0$ is small, the algebraic count of zeroes of F_j on the region $\{|s| \geq s_0\}$ is thus

$$\pm \left[\mathrm{wind}^\tau\left(F_j(\pm\infty, \cdot)\right) - \mathrm{wind}^\tau\left(F_j(\pm s_0, \cdot)\right) \right]$$
$$= \mp\, \mathrm{wind}^\tau\left(h_u(\pm s_0, \cdot + j/k) - h_v(\pm s_0, \cdot)\right);$$

i.e., it is the *relative asymptotic winding number* of v about the reparametrization $u(s, t+j/k)$, with respect to the trivialization τ. Summing this over all such reparametrizations gives

$$\sum_{j=0}^{k-1} \mp\, \mathrm{wind}^\tau\left(h_u(s, \cdot + j/k) - h_v(s, \cdot)\right), \tag{4.3}$$

4.2 Relative Intersection Numbers and the ∗-Pairing 69

where the parameter s can be chosen to be any number sufficiently close to $\pm\infty$. If $k_z \neq k_\zeta$, then the above computation is valid for the covers $u^{k_\zeta}(s,t) := u(k_\zeta s, k_\zeta t)$ and $v^{k_z}(s,t) := v(k_z s, k_z t)$, both asymptotic to $\gamma^{k_z k_\zeta}$, and (4.3) must then be divided by $k_z k_\zeta$ to compute $\iota_\infty^\tau(u,z;v,\zeta)$.

Remark 4.10 The computation above can be interpreted in terms of braids: namely, if u and v have at most finitely many intersections, then their asymptotic behavior at a pair of punctures with matching asymptotic orbit (up to multiplicity) determines up to isotopy a pair of (perhaps multiply covered) disjoint connected braids, whose linking number with each other is (up to a sign) $\iota_\infty^\tau(u,z;v,\zeta)$. It appears in this form in the work of Hutchings on embedded contact homology (see Appendix C.6 for further discussion).

The discussion thus far has been valid for any pair of asymptotically cylindrical maps. If we now assume u and v are also J-holomorphic, then Theorem 3.12 expresses the summands in (4.3) as winding numbers of certain relative asymptotic eigenfunctions for $\gamma^{k_z k_\zeta}$, and these winding numbers satisfy a priori bounds due to Theorem 3.15. Specifically, assume $u(s,t)$ and $v(s,t)$ approach their respective covers of γ along asymptotic eigenfunctions f_u and f_v with decay rates $|\lambda_u|$ and $|\lambda_v|$, respectively, so by (3.9) we have

$$\mp \operatorname{wind}^\tau(f_u) \geq \mp \alpha_\mp^\tau(\gamma^{k_z}), \qquad \mp \operatorname{wind}^\tau(f_v) \geq \mp \alpha_\mp^\tau(\gamma^{k_\zeta}).$$

Then the covers $u^{k_\zeta}(s,t)$ and $v^{k_z}(s,t)$ approach $\gamma^{k_z k_\zeta}$ along asymptotic eigenfunctions $f_u^{k_\zeta}$ and $f_v^{k_z}$ with decay rates $k_\zeta|\lambda_u|$ and $k_z|\lambda_v|$, respectively, and the winding is bounded by

$$\mp \operatorname{wind}^\tau(f_u^{k_\zeta}) \geq \mp k_\zeta \alpha_\mp^\tau(\gamma^{k_z}), \qquad \mp \operatorname{wind}^\tau(f_v^{k_z}) \geq \mp k_z \alpha_\mp^\tau(\gamma^{k_\zeta}).$$

Theorem 3.12 now implies that the relative decay rate controlling the approach of $v(s,t)$ to any of the reparametrizations $u(s, t + j/k)$ is at least the minimum of $k_\zeta|\lambda_u|$ and $k_z|\lambda_v|$, and thus the corresponding winding number is similarly bounded due to Theorem 3.15. We conclude that each of the summands in (4.3) is bounded from below by the integer $\Omega_\pm^\tau(\gamma^{k_z}, \gamma^{k_\zeta})$, where for any $k, m \in \mathbb{N}$ we define

$$\Omega_\pm^\tau(\gamma^k, \gamma^m) := \min\left\{\mp k\alpha_\mp^\tau(\gamma^m), \mp m\alpha_\mp^\tau(\gamma^k)\right\}. \tag{4.4}$$

Adding the summands in (4.3) for $j = 0, \ldots, k_z k_\zeta - 1$ and then dividing by the combinatorial factor $k_z k_\zeta$ produces the bound

$$\iota_\infty^\tau(u,z;v,\zeta) \geq \Omega_\pm^\tau(\gamma_z^{k_z}, \gamma_\zeta^{k_\zeta}) \quad \text{if } \gamma_z = \gamma_\zeta.$$

If we extend the definition of $\Omega_\pm^\tau$ by setting

$$\Omega_\pm^\tau(\gamma_1^k, \gamma_2^m) := 0 \quad \text{whenever } \gamma_1 \neq \gamma_2,$$

70 *Intersection Theory for Punctured Holomorphic Curves*

then a universal lower bound for $\iota_\infty^\tau(u, v)$ can now be written in terms of asymptotic winding numbers as

$$\iota_\infty^\tau(u, v) \geq \sum_{(z,\zeta)\in\Gamma_u^\pm\times\Gamma_v^\pm} \Omega_\pm^\tau(\gamma_z^{k_z}, \gamma_\zeta^{k_\zeta}). \tag{4.5}$$

Definition 4.11 For any asymptotically cylindrical maps $u: \dot{\Sigma} \to \widehat{W}$ and $v: \dot{\Sigma}' \to \widehat{W}$ with finitely many intersections, define

$$\iota_\infty(u, v) := \iota_\infty^\tau(u, v) - \sum_{(z,\zeta)\in\Gamma_u^\pm\times\Gamma_v^\pm} \Omega_\pm^\tau(\gamma_z^{k_z}, \gamma_\zeta^{k_\zeta}).$$

Similarly, for *any* asymptotically cylindrical maps u and v (not necessarily with finitely many intersections), we can define

$$u * v := u \bullet_\tau v - \sum_{(z,\zeta)\in\Gamma_u^\pm\times\Gamma_v^\pm} \Omega_\pm^\tau(\gamma_z^{k_z}, \gamma_\zeta^{k_\zeta}).$$

When it is well defined, $\iota_\infty(u, v)$ is sometimes called the **asymptotic contribution** to $u * v$.

Exercise 4.12 Check that neither of the above definitions depends on the choice of trivializations τ.

Definitions involving $\Omega_\pm^\tau(\gamma^k, \gamma^m)$ may seem not very enlightening at first, and they are seldom used in practice for computations, but it's useful to keep in mind what these terms mean: they are *theoretical bounds* on the possible relative asymptotic winding of ends of u around (all possible reparametrizations of) ends of v. We will say that a given winding number is **extremal** whenever it achieves the corresponding theoretical bound. We conclude, for example:

Theorem 4.13 (*asymptotic positivity of intersections*) *If u and v are asymptotically cylindrical J-holomorphic curves with nonidentical images, then $\iota_\infty(u, v) \geq 0$, with equality if and only if for all pairs of ends of u and v respectively asymptotic to covers of the same Reeb orbit, all of the resulting relative asymptotic eigenfunctions have extremal winding.* □

It is also immediate from the above definition that $u * v$ is homotopy invariant and equals $u \cdot v + \iota_\infty(u, v)$ whenever $u \cdot v$ is well defined. This completes the proof of Theorem 4.1, except for the claim that one can always achieve $u \cdot v = u * v$ after a perturbation of the data. This can be proved by observing that the subset of $\mathcal{M}_g(\widehat{W}, J) \times \mathcal{M}_{g'}(\widehat{W}, J)$ consisting of pairs (u, v) with $\iota_\infty(u, v) > 0$ consists precisely of those pairs that share an asymptotic orbit at which some relative asymptotic winding number is not extremal. By Theorem 3.15, this means that the relative decay rate of some pair of ends approaching the same

4.3 Adjunction Formulas, Relative and Absolute

orbit is an eigenvalue other than the one closest to 0. One can then use Fredholm theory with exponential weights (cf. [HWZ99, Wen10a, Hry12]) to show that the moduli space of pairs of curves satisfying this relative decay condition has strictly smaller Fredholm index than the usual moduli space, and thus for generic data, it is a submanifold of positive codimension.

4.3 Adjunction Formulas, Relative and Absolute

In order to generalize the adjunction formula, we begin by computing $u \bullet_\tau u$ for an immersed simple J-holomorphic curve $u: (\dot{\Sigma}, j) \to (\widehat{W}, J)$ with asymptotic orbits

$$\{\gamma_z^{k_z}\}_{z \in \Gamma^\pm},$$

where the notation is chosen, as in the previous section, so that γ_z is always a simply covered orbit and $k_z \in \mathbb{N}$ is the corresponding covering multiplicity. Choose a section η of the normal bundle $N_u \to \dot{\Sigma}$ with finitely many zeroes, such that, on each cylindrical neighborhood $Z_\pm \subset \dot{\Sigma}$ of a puncture $z \in \Gamma^\pm$,

$$\eta(s, t) \to \eta_\infty(k_z t) \quad \text{uniformly in } t \text{ as } s \to \pm\infty,$$

for some nonzero $\eta_\infty \in \Gamma(\gamma_z^* \xi_\pm)$ satisfying $\text{wind}^\tau(\eta_\infty) = 0$. We can also assume the zeroes of η are disjoint from all points $z \in \dot{\Sigma}$ at which $u(z) = u(\zeta)$ for some $\zeta \neq z$. Then $u \bullet_\tau u = u \cdot u^\tau$, where

$$u^\tau(z) = \exp_{u(z)} \epsilon \eta(z)$$

for some $\epsilon > 0$ small. As we saw in §2.1, there are two obvious sources of intersections between u and u^τ:

(1) Each transverse double point $u(z) = u(\zeta)$ with $z \neq \zeta$ contributes two transverse positive intersections, one near z and one near ζ. More generally, the algebraic count of intersections contributed by each isolated double point is twice its local intersection index.

(2) Each zero $\eta(z) = 0$ contributes an intersection at z, with local intersection index equal to the order of the zero. The algebraic count of these zeroes is the relative first Chern number $c_1^\tau(N_u) \in \mathbb{Z}$.

Unlike in the closed case, there are now two additional sources of intersections. As we saw in the previous section, if $z, \zeta \in \Gamma^\pm$ are two distinct punctures with $\gamma_z = \gamma_\zeta$, then perturbing u to u^τ will cause $\iota_\infty^\tau(u, z; u, \zeta) \in \mathbb{Z}$ additional intersections of u and u^τ to appear near infinity along the corresponding half-cylinders, and this number is also bounded below by $\Omega_\pm^\tau(\gamma_z^{k_z}, \gamma_\zeta^{k_\zeta})$, defined

72   *Intersection Theory for Punctured Holomorphic Curves*

in (4.4) in terms of winding numbers of asymptotic eigenfunctions. Additionally, near any $z \in \Gamma^\pm$ with $k_z > 1$, u may intersect different parametrizations of u^τ near infinity. To see this, we can again parametrize a neighborhood of z in $\dot{\Sigma}$ with the half-cylinder $Z_\pm$ and write

$$u(s, t) = \exp_{(kTs, \gamma(kt))} h(s, t)$$

for large $|s|$, where $k := k_z$, $\gamma := \gamma_z$, $T > 0$ is the period of γ and $h(s, t) \in \xi_\pm$ decays to 0 as $s \to \pm\infty$. Since u is simple, it has no double points in some neighborhood of infinity, which means that for some $s_0 > 0$, we have

$$h(s, t) \neq h(s, t + j/k) \quad \text{for all } |s| \geq s_0, t \in S^1, j \in \{1, \ldots, k - 1\}.$$

The perturbation u^τ on this neighborhood may be assumed to take the form

$$u^\tau(s, t) = \exp_{(kTs, \gamma(kt))} \left[h(s, t) + \epsilon \eta_\infty(kt) \right],$$

for some $\epsilon > 0$ small, where again $\mathrm{wind}^\tau(\eta_\infty) = 0$. Thus intersections of u with u^τ on the region $\{|s| \geq s_0\}$ correspond to solutions of

$$F_j(s, t) := h(s, t + j/k) - h(s, t) - \epsilon \eta_\infty(kt) = 0$$

for arbitrary values of $j = 0, \ldots, k - 1$. For $j = 0$, this equation has no solutions. For $j = 1, \ldots, k - 1$, we observe that F_j extends continuously to $s = \pm\infty$ with $F_j(\pm\infty, t) = -\epsilon \eta_\infty(kt)$ and obtain the count of solutions

$$\pm \left[\mathrm{wind}^\tau \left(F_j(\pm\infty, \cdot) \right) - \mathrm{wind}^\tau \left(F_j(\pm s_0, \cdot) \right) \right]$$
$$= \mp \mathrm{wind}^\tau \left(h(\pm s_0, \cdot + j/k) - h(\pm s_0, \cdot) \right).$$

The count of additional intersections of u with u^τ in a neighborhood of z is therefore

$$\iota_\infty^\tau(u, z) := \mp \sum_{j=1}^{k_z - 1} \mathrm{wind}^\tau \left(h(s, \cdot + j/k) - h(s, \cdot) \right), \tag{4.6}$$

where s is any number sufficiently close to $\pm\infty$, and we can then write the total count of asymptotic contributions to $u \bullet_\tau u$ as

$$\iota_\infty^\tau(u) := \sum_{z, \zeta \in \Gamma^\pm, z \neq \zeta} \iota_\infty^\tau(u, z; u, \zeta) + \sum_{z \in \Gamma^\pm} \iota_\infty^\tau(u, z).$$

This yields the computation

$$u \bullet_\tau u = 2\delta(u) + c_1^\tau(N_u) + \iota_\infty^\tau(u),$$

and, since $c_1^\tau(N_u) = c_1^\tau(u^* T \widehat{W}) - \chi(\dot{\Sigma})$, we deduce from this a relation that is valid for any (not necessarily immersed) simple and asymptotically cylindrical J-holomorphic curve, called the **relative adjunction formula**

4.3 Adjunction Formulas, Relative and Absolute

$$u \bullet_\tau u = 2\delta(u) + c_1^\tau(u^* T\widehat{W}) - \chi(\dot{\Sigma}) + \iota_\infty^\tau(u). \tag{4.7}$$

This version of the adjunction formula first appeared in [Hut02].

Remark 4.14 As with $\iota_\infty^\tau(u, z\,;\, v, \zeta)$ (cf. Remark 4.10), $\iota_\infty^\tau(u, z)$ can be given a braid-theoretic interpretation: it is (up to a sign) the writhe of the braid defined by identifying a neighborhood of the framed loop γ_z with $S^1 \times \mathbb{D}$ and projecting the embedded loop $u(s, \cdot)$ to $M_\pm$ for any s close to $\pm\infty$ (see Appendix C.6).

As we did with $\iota_\infty^\tau(u, v)$ in the previous section, it will be useful to derive a theoretical bound on $\iota_\infty^\tau(u)$. We already have $\iota_\infty^\tau(u, z\,;\, u, \zeta) \geq \Omega_\pm^\tau(\gamma_z^{k_z}, \gamma_\zeta^{k_\zeta})$, and must deduce a similar bound for $\iota_\infty^\tau(u, z)$. Let $\gamma := \gamma_z$ and $k := k_z$, and write $u(s, t) = \exp_{(Ts, \gamma(kt))} h(s, t)$ as usual, and, for $j = 1, \ldots, k - 1$, let

$$u_j(s, t) := u(s, t + j/k) = \exp_{(Ts, \gamma(kt))} h_j(s, t), \quad \text{where } h_j(s, t) := h(s, t + j/k).$$

By Theorem 3.11, $h(s, t)$ is controlled as $s \to \pm\infty$ by some eigenfunction f of $\mathbf{A}_{\gamma^k}$ with eigenvalue λ, and, by (3.9),

$$\mp \operatorname{wind}^\tau(f) \geq \mp\alpha_\mp^\tau(\gamma^k).$$

The reparametrizations $h_j(s, t)$ are similarly controlled by reparametrized eigenfunctions

$$f_j(t) := f(t + j/k)$$

with $\operatorname{wind}^\tau(f_j) = \operatorname{wind}^\tau(f)$ and the same eigenvalue, and the relative decay rates controlling $h_j - h$ are then at least $|\lambda|$ due to Theorem 3.12, implying (via Theorem 3.15) a corresponding bound on the relative winding terms in (4.6); thus

$$\iota_\infty^\tau(u, z) \geq \mp(k - 1) \operatorname{wind}^\tau(f) \geq \mp(k - 1)\alpha_\mp^\tau(\gamma^k).$$

The bound established above is only a first attempt, as we will see in a moment that a stricter bound may hold in general. If $\operatorname{wind}^\tau(f)$ is not extremal, i.e., $\mp \operatorname{wind}^\tau(f) \geq \mp\alpha_\mp^\tau(\gamma^k) + 1$, the above computation gives

$$\iota_\infty^\tau(u, z) \geq \mp(k - 1)\alpha_\mp^\tau(\gamma^k) + k - 1. \tag{4.8}$$

Alternatively, suppose $\operatorname{wind}^\tau(f)$ is extremal, hence equal to $\alpha_\mp^\tau(\gamma^k)$, and let $m = \operatorname{cov}(f)$, so $k = m\ell$ for some $\ell \in \mathbb{N}$ and

$$f(t) = g^m(t) := g(mt)$$

for some eigenfunction g of $\mathbf{A}_{\gamma^\ell}$ that is simply covered. It follows that for $j = 1, \ldots, k - 1$, $f_j \equiv f$ if and only if $j \in \ell\mathbb{Z}$. When j is not divisible by ℓ, Theorem 3.12 now gives a relative decay rate equal to $|\lambda|$ and thus relative

74 *Intersection Theory for Punctured Holomorphic Curves*

winding equal to $\operatorname{wind}^\tau(f)$, so adding up these terms for the $m(\ell - 1)$ values of j not in $\ell\mathbb{Z}$ contributes

$$\mp m(\ell - 1)\alpha^\tau_\mp(\gamma^k) \tag{4.9}$$

to $\iota^\tau_\infty(u, z)$.

For $j = 1, \ldots, m - 1$, we claim that the asymptotic winding of $h_{j\ell} - h$ is stricter than the bound established above; i.e., for large $|s|$,

$$\mp \operatorname{wind}^\tau\left(h_{j\ell}(s, \cdot) - h(s, \cdot)\right) \geq \mp\alpha^\tau_\mp(\gamma^k) + 1. \tag{4.10}$$

By Theorem 3.12, there is a nontrivial eigenfunction $\varphi_j \in \Gamma((\gamma^k)^*\xi_\pm)$ of $\mathbf{A}_{\gamma^k}$ with eigenvalue λ', such that

$$h_{j\ell}(s, t) - h(s, t) = e^{\lambda's}\left[\varphi_j(t) + r'(s, t)\right],$$

for large $|s|$, with $r'(s, \cdot) \to 0$ uniformly as $s \to \pm\infty$. Now if the claim is false, then $\operatorname{wind}^\tau(\varphi_j) = \alpha^\tau_\mp(\gamma^k) = \operatorname{wind}^\tau(f)$. Since f is an m-fold cover, this means $\operatorname{wind}^\tau(\varphi_j)$ is divisible by m, and Remark 4.3 then implies that φ_j is also an m-fold cover, and thus

$$\varphi_j(t + 1/m) = \varphi_j(t) \quad \text{for all } t \in S^1. \tag{4.11}$$

But observe:

$$\begin{aligned}
0 &= \sum_{r=0}^{m-1}\left[h\left(s, t + \frac{j+r}{m}\right) - h\left(s, t + \frac{r}{m}\right)\right] \\
&= \sum_{r=0}^{m-1}\left[h_{j\ell}\left(s, t + \frac{r}{m}\right) - h\left(s, t + \frac{r}{m}\right)\right] \\
&= \sum_{r=0}^{m-1}e^{\lambda's}\left[\varphi_j\left(t + \frac{r}{m}\right) + r'\left(s, t + \frac{r}{m}\right)\right],
\end{aligned}$$

implying

$$\sum_{r=0}^{m-1}\varphi_j(t + r/m) = 0 \quad \text{for all } t \in S^1.$$

Since φ_j is not identically zero, this contradicts (4.11) and thus proves the claim. Adding to (4.9) the $m - 1$ terms bounded by (4.10), we conclude

$$\begin{aligned}
\iota^\tau_\infty(u, z) &\geq \mp m(\ell - 1)\alpha^\tau_\mp(\gamma^k) + (m - 1)\left[\mp\alpha^\tau_\mp(\gamma^k) + 1\right] \\
&= \mp(k - 1)\alpha^\tau_\mp(\gamma^k) + (m - 1).
\end{aligned}$$

This bound is weaker than (4.8), but the latter is valid only when $\operatorname{wind}^\tau(f)$ is nonextremal; thus the former is the strongest possible bound in general. Recall

4.3 Adjunction Formulas, Relative and Absolute

that the covering multiplicity $m = \mathrm{cov}(f)$ is precisely what we denoted by $\bar{\sigma}_{\mp}(\gamma^k)$ in §4.1. To summarize, we now define, for any simply covered orbit γ and $k \in \mathbb{N}$,

$$\Omega_{\pm}^{\tau}(\gamma^k) := \mp(k-1)\alpha_{\mp}^{\tau}(\gamma^k) + \left[\bar{\sigma}_{\mp}(\gamma^k) - 1\right]. \tag{4.12}$$

The above computation then implies

$$\iota_{\infty}^{\tau}(u, z) \geq \Omega_{\pm}^{\tau}(\gamma_z^{k_z}). \tag{4.13}$$

Definition 4.15 For any asymptotically cylindrical map $u \colon \dot{\Sigma} \to \widehat{W}$ that is embedded outside some compact set, we define the **asymptotic contribution** to the singularity index by

$$\delta_{\infty}(u) := \frac{1}{2}\left[\iota_{\infty}^{\tau}(u) - \sum_{z,\zeta \in \Gamma^{\pm}, z \neq \zeta} \Omega_{\pm}^{\tau}(\gamma_z^{k_z}, \gamma_\zeta^{k_\zeta}) - \sum_{z \in \Gamma^{\pm}} \Omega_{\pm}^{\tau}(\gamma_z^{k_z})\right].$$

Exercise 4.16 Check that the above definition does not depend on the trivializations τ. Then try to convince yourself that it's an integer, not a half-integer.

Like Theorem 4.13 in the previous section, the following is now immediate from the computation above:

Theorem 4.17 *If u is an asymptotically cylindrical and simple J-holomorphic curve, then $\delta_{\infty}(u) \geq 0$, with equality if and only if:*

(1) *For all pairs of ends asymptotic to covers of the same Reeb orbit, the resulting relative asymptotic eigenfunctions have extremal winding.*
(2) *For all ends asymptotic to multiply covered Reeb orbits, the relative asymptotic eigenfunctions controlling the approach of distinct branches to each other have extremal winding.*

The proof of the absolute adjunction formula in Theorem 4.4 now consists only of plugging in the relevant definitions and computing.

Exercise 4.18 Show that for any simply covered Reeb orbit γ and $k \in \mathbb{N}$,

$$\Omega_{\pm}^{\tau}(\gamma^k) - \Omega_{\pm}^{\tau}(\gamma^k, \gamma^k) \mp \alpha_{\mp}^{\tau}(\gamma^k) = \bar{\sigma}_{\mp}(\gamma^k) - 1.$$

Proof of Theorem 4.4 Plugging the relative adjunction formula (4.7) into the definition of $u * u$ (Definition 4.11) gives

Intersection Theory for Punctured Holomorphic Curves

$$u * u = u \bullet_\tau u - \sum_{(z,\zeta) \in \Gamma^\pm \times \Gamma^\pm} \Omega_\pm^\tau(\gamma_z^{k_z}, \gamma_z^{k_z})$$

$$= 2\delta(u) + c_1^\tau(u^*T\widehat{W}) - \chi(\dot{\Sigma}) + \iota_\infty^\tau(u) - \sum_{(z,\zeta) \in \Gamma^\pm \times \Gamma^\pm} \Omega_\pm^\tau(\gamma_z^{k_z}, \gamma_z^{k_z}).$$

Now replacing $c_1^\tau(u^*T\widehat{W}) - \chi(\dot{\Sigma})$ with $c_N(u)$ plus some extra terms from Definition 3.22, and $\iota_\infty^\tau(u)$ with $2\delta_\infty(u)$ plus extra terms from Definition 4.15, all terms of the form $\Omega_\pm^\tau(\gamma_z^{k_z}, \gamma_\zeta^{k_\zeta})$ with $z \neq \zeta$ cancel and the above becomes

$$u * u = 2\left[\delta(u) + \delta_\infty(u)\right] + c_N(u) + \sum_{z \in \Gamma^\pm} \left[\Omega_\pm^\tau(\gamma_z^{k_z}) - \Omega_\pm^\tau(\gamma_z^{k_z}, \gamma_z^{k_z}) \mp \alpha_\mp^\tau(\gamma_z^{k_z})\right].$$

The result then follows from Exercise 4.18. $\qquad\square$

Exercise 4.19 Assume $\gamma \colon S^1 \to M$ is a nondegenerate Reeb orbit in a contact 3-manifold $(M, \xi = \ker \alpha)$, and given $J \in \mathcal{J}(\alpha)$, let $u_\gamma \colon \mathbb{R} \times S^1 \to \mathbb{R} \times M$ denote the associated J-holomorphic orbit cylinder.

(a) Show that $c_N(u_\gamma) = -p(\gamma)$, where $p(\gamma) \in \{0, 1\}$ is the parity of the Conley–Zehnder index of γ.

(b) Show that $u_\gamma * u_\gamma = -\text{cov}(\gamma) \cdot p(\gamma)$.

(c) Deduce from part (b) that if u^k denotes a k-fold cover of a given asymptotically cylindrical J-holomorphic curve u, it is *not* generally true that $u^k * v^\ell = k\ell(u * v)$.

   Remark: One can show however that, in general, $u^k * v^\ell \geq k\ell(u * v)$ (cf. Proposition C.2).

(d) Use the adjunction formula to show the following: if γ is a multiple cover of a Reeb orbit with even Conley–Zehnder index, and J' is an arbitrary almost complex structure on $\mathbb{R} \times M$ that is compatible with $d(e^s\alpha)$ and belongs to $\mathcal{J}(\alpha)$ outside a compact subset, then there is no simple J'-holomorphic curve homotopic to u_γ through asymptotically cylindrical maps.

Lecture 5

Symplectic Fillings of Planar Contact 3-Manifolds

In this lecture, we will explain an application of the intersection theory of punctured holomorphic curves to the problem of classifying symplectic fillings of contact 3-manifolds. The main result is stated in §5.2 as Theorem 5.6, and it may be seen as an analogue of McDuff's Theorem 1.16 in a slightly different context – indeed, the structure of the proof is very similar, but the technical details are a bit more intricate and require the machinery developed in Lecture 4. Before stating the theorem and sketching its proof, we review some topological facts about Lefschetz fibrations, open books, and symplectic fillings.

5.1 Open Books and Lefschetz Fibrations

As we saw in Lecture 1, symplectic forms on 4-manifolds can be characterized topologically (up to deformation) via Lefschetz fibrations. The natural analogue of a Lefschetz fibration for a contact manifold is an **open book decomposition**. If M is a closed oriented 3-manifold, an open book is a pair (B, π), where $B \subset M$ is an oriented link and

$$\pi \colon M \backslash B \to S^1$$

is a fibration such that some neighborhood $\mathcal{N}(\gamma) \subset M$ of each connected component $\gamma \subset B$ admits an identification with $S^1 \times \mathbb{D}$ in which π takes the form

$$\pi|_{\mathcal{N}(\gamma)\backslash\gamma} \colon S^1 \times (\mathbb{D}\backslash\{0\}) \to S^1 \colon (\theta, (r, \phi)) \mapsto \phi.$$

Here (r, ϕ) denote polar coordinates on the disk $\mathbb{D}$, with the angle normalized to take values in $S^1 = \mathbb{R}/\mathbb{Z}$. We call B the **binding** of the open book, and

Symplectic Fillings of Planar Contact 3-Manifolds

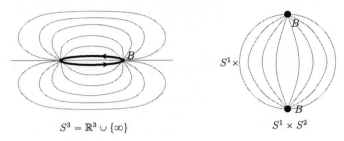

Figure 5.1 Simple open book decompositions on S^3 and $S^1 \times S^2$, with pages diffeomorphic to the plane and the cylinder, respectively.

the fibers $\pi^{-1}(\phi) \subset M$ are its **pages**; these are open surfaces whose closures are compact oriented surfaces with oriented boundary equal to B. Figure 5.1 shows simple examples on S^3 and $S^1 \times S^2$.

A contact structure ξ on M is said to be **supported** by the open book $\pi \colon M \backslash B \to S^1$ if one can write $\xi = \ker \alpha$ for some contact form α with

$$\alpha|_{TB} > 0 \quad \text{and} \quad d\alpha|_{\text{pages}} > 0.$$

Equivalently, one can require that the components of B are closed Reeb orbits with respect to α, and everywhere else the Reeb vector field is positively transverse to the pages. This definition is due to Giroux, and contact forms that satisfy these conditions are sometimes called **Giroux forms**.

The following contact analogue of Theorems 1.3 and 1.8 is a translation into modern language of a classical result of Thurston and Winkelnkemper:

Theorem 5.1 (Thurston and Winkelnkemper [TW75]) *Every open book on a closed oriented 3-manifold supports a unique contact structure up to isotopy.*

A much deeper result, known as the *Giroux correspondence* [Gir02], asserts that the set of contact structures up to isotopy on any closed 3-manifold admits a natural bijection to the set of open books up to a topological operation called *positive stabilization*. We will not need this fact in the discussion below, but it is worth mentioning since it has had a major impact on the modern field of contact topology (see, e.g., [Etn06] for more on this subject).

In order to discuss symplectic fillings, we will also need to consider a more general class of Lefschetz fibrations, in which both the base and fiber are allowed to have boundary. Specifically, assume W is a compact oriented 4-manifold with boundary and corners, where the boundary consists of two smooth faces

$$\partial W = \partial_h W \cup \partial_v W,$$

5.1 Open Books and Lefschetz Fibrations

the **horizontal** and **vertical boundary**, respectively, which intersect each other at a corner of codimension 2. Given a compact oriented surface Σ with nonempty boundary, we define a **bordered Lefschetz fibration** of W over Σ to be a smooth map

$$\Pi : W \to \Sigma$$

with finitely many *interior* critical points $W^{\mathrm{crit}} := \mathrm{Crit}(\Pi) \subset \mathring{W}$ and critical values $\Sigma^{\mathrm{crit}} := \Pi(W^{\mathrm{crit}}) \subset \mathring{\Sigma}$ such that:

(1) As in Example 1.5, critical points take the form $\Pi(z_1, z_2) = z_1^2 + z_2^2$ in complex local coordindates compatible with the orientations.
(2) The fibers have nonempty boundary.
(3) $\Pi^{-1}(\partial\Sigma) = \partial_v W$, and

$$\Pi|_{\partial_v W} : \partial_v W \to \partial\Sigma$$

   is a smooth fibration.
(4) $\partial_h W = \bigcup_{z \in \Sigma} \partial\left(\Pi^{-1}(z)\right)$, and

$$\Pi|_{\partial_h W} : \partial_h W \to \Sigma$$

   is also a smooth fibration.

In the following, we assume the base is the closed unit disk (see Figure 5.2),

$$\Sigma := \mathbb{D} \subset \mathbb{C}.$$

In this case, the vertical boundary is a connected fibration of some compact oriented surface with boundary over $\partial\mathbb{D} = S^1$,

$$\pi := \Pi|_{\partial_v W} : \partial_v W \to S^1,$$

and the horizontal boundary is a disjoint union of circle bundles over $\mathbb{D}$; since bundles over $\mathbb{D}$ are trivial, the connected components of $\partial_h W$ can then be identified with $S^1 \times \mathbb{D}$ such that π on the corner $\partial_h W \cap \partial_v W = \partial(\partial_h W) = \coprod(S^1 \times \partial\mathbb{D})$ takes the form $\pi(\theta, \phi) = \phi$. This means that after smoothing the corners of ∂W, the latter inherits from $\Pi : W \to \mathbb{D}$ an open book decomposition $\pi : \partial W \backslash B \to S^1$ uniquely up to isotopy, with $\partial_h W$ regarded as a tubular neighborhood of the binding $B := \coprod(S^1 \times \{0\})$.

Recall that for any surface fibration $F \hookrightarrow M \to S^1$ that is trivial near the boundaries of the fibers, the parallel transport (with respect to any connection) along a full traversal of the loop S^1 determines (uniquely up to isotopy) a diffeomorphism $\varphi : F \to F$ that is trivial near ∂F; we call this the **monodromy** of the fibration. One can thus define the monodromy of a Lefschetz fibration along any loop containing no critical values – in particular, the monodromy along $\partial\mathbb{D}$ is also called the *monodromy of the open book* induced at the boundary.

Symplectic Fillings of Planar Contact 3-Manifolds

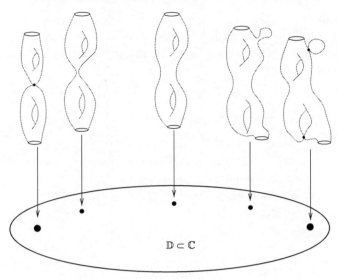

Figure 5.2 A bordered Lefschetz fibration over the unit disk $\mathbb{D} \subset \mathbb{C}$, where the regular fibers have genus 2 and two boundary components, and there are two singular fibers, each with two irreducible components. The boundary inherits an open book with pages of genus 2 and two binding components.

It is a basic fact about the topology of Lefschetz fibrations that the monodromy along a loop can always be expressed in terms of *positive Dehn twists* (see, e.g., [GS99]). For our purposes, the relevant version of this statement is the following. Let $z_0 = 1 \in \partial \mathbb{D}$ and denote the fiber at z_0 by $F := \Pi^{-1}(z_0)$. Pick a set of smooth paths

$$\gamma_z \colon [0, 1] \to \mathbb{D}, \quad \text{for each } z \in \mathbb{D}^{\mathrm{crit}},$$

from $\gamma_z(0) = z_0$ to $\gamma_z(1) = z$, intersecting each other only at z_0. Then, for each $z \in \mathbb{D}^{\mathrm{crit}}$ and $p \in W^{\mathrm{crit}} \cap \Pi^{-1}(z)$, there is a unique isotopy class of smoothly embedded circles

$$S^1 \cong C_p \subset F$$

that can be collapsed to p under parallel transport along γ_z; this is called the **vanishing cycle** of p. We then have the following proposition:

Proposition 5.2 *If $\Pi \colon W \to \mathbb{D}$ is a bordered Lefschetz fibration, then the monodromy $F \to F$ of the induced open book at the boundary is a composition of positive Dehn twists along the vanishing cycles $C_p \subset F$ for each critical point $p \in W^{\mathrm{crit}}$.*

5.1 Open Books and Lefschetz Fibrations

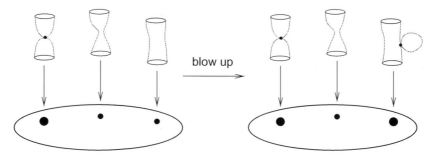

Figure 5.3 Two bordered Lefschetz fibrations with regular fiber diffeomorphic to the annulus. The picture at the right is obtained from the one at the left by blowing up at a regular point.

Example 5.3 Suppose $\Pi\colon W \to \mathbb{D}$ has regular fiber $F \cong [-1, 1] \times S^1$ and exactly $k \geq 0$ singular fibers, each consisting of two disks connected along a critical point (see Figure 5.3, left). The resulting open book on ∂W then has pages diffeomorphic to $\mathbb{R} \times S^1$ and monodromy δ^k, where δ denotes the positive Dehn twist along the separating curve $\{0\} \times S^1$, which generates the mapping class group of $\mathbb{R} \times S^1$. If we blow up W at a regular point in the interior, then by Exercise 1.11 we obtain a new bordered Lefschetz fibration with one additional singular fiber consisting of an annulus connected to an exceptional sphere (Figure 5.3, right). This blowup operation obviously does not change the open book on ∂W, which is consistent with Proposition 5.2 since the additional Dehn twist introduced by the extra critical point is along a *contractible* vanishing cycle, and is therefore isotopic to the identity.

Let us denote the contact manifolds supported by the open books on ∂W in Example 5.3 by (M_k, ξ_k). It is not too hard to say precisely what these contact manifolds are: topologically, we have $M_0 \cong S^1 \times S^2$, $M_1 \cong S^3$, and M_k for $k \geq 2$ is the lens space $L(k, k - 1)$. All of these carry *standard* contact structures that can be defined as follows: we defined (S^3, ξ_{std}) already in §2.3 by identifying S^3 with the boundary of the unit ball in $\mathbb{R}^4$ with coordinates (x_1, y_1, x_2, y_2) and writing $\xi_{\text{std}} = \ker(\lambda_{\text{std}}|_{TS^3}) \subset TS^3$, where

$$\lambda_{\text{std}} := \frac{1}{2} \sum_{j=1}^{2} \left(x_j\, dy_j - y_j\, dx_j \right).$$

Under the natural identification $\mathbb{R}^4 \to \mathbb{C}^2 \colon (x_1, y_1, x_2, y_2) \mapsto (x_1 + iy_1, x_2 + iy_2)$, this contact structure is invariant under the action of U(2); thus the standard contact structure ξ_{std} on any lens space $L(p, q)$ for two coprime integers $p > q \geq 1$ can be defined via the quotient

82   *Symplectic Fillings of Planar Contact 3-Manifolds*

$$(L(p,q), \xi_{\mathrm{std}}) := (S^3, \xi_{\mathrm{std}})/G_{p,q},$$

where $G_{p,q} \subset U(2)$ denotes the subgroup

$$G_{p,q} := \left\{ \begin{pmatrix} \zeta & 0 \\ 0 & \zeta^q \end{pmatrix} \in U(2) \,\middle|\, \zeta^p = 1 \right\}.$$

Finally, on $S^1 \times S^2$, we use the coordinates (η, θ, ϕ), where $\eta \in S^1 = \mathbb{R}/\mathbb{Z}$ and $(\theta, \phi) \in [0, \pi] \times (\mathbb{R}/2\pi\mathbb{Z})$ are the natural spherical coordinates on S^2, and write

$$\xi_{\mathrm{std}} = \ker [f(\theta)\, d\eta + g(\theta)\, d\phi]$$

for a suitably chosen loop $(f, g) \colon \mathbb{R}/\pi\mathbb{Z} \to \mathbb{R}^2 \backslash \{0\}$ that is based at the point $(f(0), g(0)) = (1, 0)$ and winds exactly once counterclockwise around the origin. Any two choices of (f, g) that make the above expression a smooth contact form on $S^1 \times S^2$ and have the stated winding property produce isotopic contact structures (see, e.g., [Gei08]). The following proposition can now be verified by constructing explicit open books that support these contact structures and then computing the monodromy.

Proposition 5.4   *There are contactomorphisms* $(M_0, \xi_0) \cong (S^1 \times S^2, \xi_{\mathrm{std}})$, $(M_1, \xi_1) \cong (S^3, \xi_{\mathrm{std}})$, *and* $(M_k, \xi_k) \cong (L(k, k-1), \xi_{\mathrm{std}}))$ *for each* $k \geq 2$.

We will say that a symplectic form ω on W is **supported by** a bordered Lefschetz fibration $\Pi \colon W \to \mathbb{D}$ if the following conditions hold:

(1) Every fiber of $\Pi|_{W \backslash W^{\mathrm{crit}}} \colon W \backslash W^{\mathrm{crit}} \to \mathbb{D}$ is a symplectic submanifold.
(2) On a neighborhood of W^{crit}, ω tames some almost complex structure J that preserves the tangent spaces of the fibers.
(3) On a neighborhood of ∂W, $\omega = d\lambda$ for some 1-form λ such that $\lambda|_{T(\partial_h W)}$ and $\lambda|_{T(\partial_v W)}$ are each contact forms, and the induced Reeb vector field on $\partial_h W$ is tangent to the fibers (in the positive direction).

Observe that for the contact form λ on the smooth faces of ∂W in the above definition, $d\lambda = \omega$ is necessarily positive on the pages of the induced open book, and λ is also positive on the binding in $\partial_h W$, so that $\lambda|_{\partial W}$ satisfies a variation on the conditions for a Giroux form. The natural analogue of Theorem 1.8 in this context is the following:

Theorem 5.5   *On any bordered Lefschetz fibration* $\Pi \colon W \to \mathbb{D}$, *the space of supported symplectic forms is nonempty and connected, and the corner of* ∂W *can be smoothed so that* (W, ω) *becomes (canonically up to symplectic deformation) a symplectic filling of the contact structure supported by the induced open book at the boundary.*

5.2 A Classification Theorem for Symplectic Fillings

We note one additional detail about this construction: a symplectic form ω on W may sometimes be exact since $\partial W \neq \varnothing$, but the condition of ω being positive on fibers imposes contraints that may make this impossible. In particular, ω can never be exact if any singular fiber of $\Pi: W \to \mathbb{D}$ contains an irreducible component that is closed – this would violate Stokes's theorem. We say that a bordered Lefschetz fibration is **allowable** if no such components exist, which is equivalent to saying that all the vanishing cycles are homologically nontrivial. For example, the Lefschetz fibration in Figure 5.2 is not allowable, due to the presence of a closed irreducible component in the singular fiber at the right, but one can show that this component is an exceptional sphere, and thus an allowable Lefschetz fibration could be produced by blowing it down (cf. Exercise 1.11).

It turns out that if $\Pi: W \to \mathbb{D}$ is allowable, one can always construct ω so that it is not only exact but also arises from a *Weinstein structure*, a much more rigid notion of a symplectic filling. We will not discuss Weinstein and Stein fillings any further here (see [Etn98, OS04a, CE12]), except to mention the following related result:

Theorem 5.5′ *If $\Pi: W \to \mathbb{D}$ is an allowable bordered Lefschetz fibration, then (W, ω) in Theorem 5.5 can be arranged to be a Weinstein filling of the contact manifold $(\partial W, \xi)$ supported by the open book induced at the boundary. In particular, $(\partial W, \xi)$ is Stein fillable.*

Theorems 5.5 and 5.5′ can be found in a variety of forms in the literature but are usually not stated quite so precisely as we have stated them here – complete proofs of our versions (including also cases where $\Sigma \neq \mathbb{D}$) may be found in [LVWa].

5.2 A Classification Theorem for Symplectic Fillings

An open book decomposition of a 3-manifold is called **planar** if its pages have genus 0, i.e., if they are punctured spheres. We then call (M, ξ) a **planar contact manifold** if M admits a planar open book supporting ξ. The planar contact manifolds play something of a special role in 3-dimensional contact topology, similar to the role of rational and ruled surfaces among symplectic 4-manifolds (see [McD90, Wen18]). It is not always easy to recognize whether a given contact structure is planar or not, but many results in either direction are known: Etnyre [Etn04] showed, for instance, that all *overtwisted* contact structures are planar, and, by an obstruction established in the same paper, the standard contact structures on unit cotangent bundles of oriented surfaces with positive genus are never planar. As we saw in Proposition 5.4, the standard

84 *Symplectic Fillings of Planar Contact 3-Manifolds*

contact structures on S^3, $S^1 \times S^2$ and $L(k, k-1)$ for $k \geq 2$ are all planar, as are all contact structures that arise on boundaries of bordered Lefschetz fibrations with genus 0 fibers.

For an arbitrary contact 3-manifold (M, ξ), the problem of classifying all of its symplectic fillings is often hopeless – many examples are known, for instance, which admit infinite (but not necessarily exhaustive) lists of pairwise nonhomeomorphic or nondiffeomorphic Stein fillings [Smi01, OS04b, AEMS08]. On the other hand, many of the earliest results on this question gave finite classifications and sometimes even *uniqueness* (up to certain obvious ambiguities) of symplectic fillings, e.g., for S^3 [Gro85, Eli90], $S^1 \times S^2$ [Eli90], the unit cotangent bundle of S^2 [Hin00], and lens spaces [McD90, Hin03, Lis08]. Most of these finiteness results can now be deduced from Theorem 5.6.

We will say that a symplectic filling (W, ω) of a contact 3-manifold (M, ξ) admits a **symplectic Lefschetz fibration over** $\mathbb{D}$ if there exists a bordered Lefschetz fibration $\Pi \colon E \to \mathbb{D}$ with a supported symplectic form ω_E such that, after smoothing the corners on ∂E, (E, ω_E) is symplectomorphic to (W, ω). Whenever this is the case, the Lefschetz fibration determines a supporting open book on (M, ξ) uniquely up to isotopy.

Theorem 5.6 ([Wen10b]) *Suppose (W, ω) is a symplectic filling of a contact 3-manifold (M, ξ) that is supported by a planar open book $\pi \colon M \backslash B \to S^1$. Then (W, ω) admits a symplectic Lefschetz fibration over $\mathbb{D}$ such that the induced open book at the boundary is isotopic to $\pi \colon M \backslash B \to S^1$. Moreover, the Lefschetz fibration is allowable if and only if (W, ω) is minimal.*

One can say slightly more: [Wen10b] shows, in fact, that the isotopy class of the Lefschetz fibration produced on (W, ω) depends only on the deformation class of the symplectic structure; hence the problem of classifying fillings up to deformation reduces to the problem of classifying Lefschetz fibrations that fill a given open book. In some cases, this provides an immediate uniqueness result. For instance:

Corollary 5.7 *The symplectic fillings of $(S^3, \xi_{\mathrm{std}})$, $(S^1 \times S^2, \xi_{\mathrm{std}})$ and $(L(k, k-1), \xi_{\mathrm{std}})$ for $k \geq 2$ are unique up to symplectic deformation equivalence and blowup.*

Proof By Proposition 5.4, the contact manifolds in question all admit supporting open books with cylindrical pages and monodromy equal to δ^k for some $k \geq 0$, where δ is the positive Dehn twist that generates the mapping class group of $\mathbb{R} \times S^1$. The only allowable Lefschetz fibration that produces such an open book at the boundary is the one with fiber $[-1, 1] \times S^1$ and exactly k

5.2 A Classification Theorem for Symplectic Fillings

singular fibers of the form pictured in Figure 5.3 at the left. Theorem 5.6 then implies that all the minimal symplectic fillings in question are supported by Lefschetz fibrations of this type, which determines their symplectic structures up to deformation equivalence via Theorem 5.5. □

Further uniqueness results along these lines have been obtained in papers by Plamenevskaya and Van Horn-Morris [PV10] and Kaloti and Li [KL16], each by studying the factorizations of mapping classes on planar surfaces into products of positive Dehn twists and then applying Theorem 5.6. In a slightly different direction, Wand [Wan12] used Theorem 5.6 to establish a new obstruction for a contact 3-manifold to be planar.

Theorem 5.5′ implies another quite general consequence of the above result:

Corollary 5.8 *Every symplectic filling of a planar contact manifold is deformation equivalent to a blowup of a Stein filling. In particular, any contact manifold that is both planar and symplectically fillable is also Stein fillable.*

Ghiggini [Ghi05] gave examples of contact 3-manifolds that are symplectically but not Stein fillable; hence Corollary 5.8 implies that Ghiggini's examples cannot be planar. Similarly, Wand [Wan15] and Baker, Etnyre and Van Horn-Morris [BEV12] have given examples of Stein fillable contact manifolds with (necessarily nonplanar) supporting open books that cannot arise from boundaries of Lefschetz fibrations.

Remark 5.9 Theorem 5.6 and Corollary 5.8 can be generalized to allow Lefschetz fibrations over arbitrary compact oriented surfaces with boundary [LVWb]. In this form, they apply to a larger class of contact manifolds, including many that are not planar; a prototypical example of this is the uniqueness (proved originally in [Wen10b]) of strong fillings of the 3-torus, whose tight contact structures are never planar. Generalizing in a different direction, [NW11] shows that both results also remain valid (but only specifically for *planar* contact manifolds) if the symplectic filling condition on (M, ξ) is weakened to the existence of a compact symplectic manifold (W, ω) with $\partial W = M$ and $\omega|_\xi > 0$. Such objects are known as **weak symplectic fillings** of (M, ξ), and they have been extensively studied, but at present, the planar contact manifolds are the only class for which any meaningful classification of weak fillings is known to be feasible.[1]

---

[1] The loophole in this statement is that by a frequently used lemma of Eliashberg [Eli91, Prop. 3.1], weak fillings for which ω is exact near the boundary can always be deformed to strong fillings, and thus whenever M happens to be a rational homolgy 3-sphere, the classification of weak fillings is the same as that of strong fillings. This is, however, an essentially topological phenomenon that has little to do with contact geometry.

86 *Symplectic Fillings of Planar Contact 3-Manifolds*

5.3 Sketch of the Proof

As in our proof of McDuff's result on ruled surfaces, the main idea for Theorem 5.6 is to consider a moduli space of holomorphic curves whose intersection theory is sufficiently well behaved to view them as fibers of a Lefschetz fibration. The first step is thus to define the moduli space and show that it is nonempty. This rests on a construction known as the *holomorphic open book*; the following result was first stated in [ACH05], and two proofs later appeared in independent work of Abbas [Abb11] and the author [Wen10c].

Theorem 5.10 (*holomorphic open book construction*) *Suppose (M, ξ) is supported by a planar open book $\pi \colon M \backslash B \to S^1$. Then there exists a Giroux form α and an almost complex structure $J_+ \in \mathcal{J}(\alpha)$ such that:*

(1) *Each connected component $\gamma \subset B$ is a nondegenerate Reeb orbit with $\mu_{\mathrm{CZ}}^\tau(\gamma) = 1$, where τ is any trivialization in which the pages approach γ with winding number 0.*

(2) *Each page $P \subset M$ lifts to an embedded asymptotically cylindrical J_+-holomorphic curve $u_P \colon \dot{\Sigma} \to \mathbb{R} \times M$ with all punctures positive, and $\mathrm{ind}(u_P) = 2$.*

Observe that pages of an open book come always in 1-parameter families, but when we lift them to the symplectization $\mathbb{R} \times M$, an additional parameter appears due to the translation invariance. Thus Theorem 5.10 produces a 2-dimensional moduli space

$$\mathcal{M}_{\mathrm{OB}}^+ \subset \mathcal{M}_0(\mathbb{R} \times M, J_+)$$

of J_+-holomorphic pages; it is diffeomorphic to $\mathbb{R} \times S^1$ and admits a free action by $\mathbb{R}$-translations, so that

$$\mathcal{M}_{\mathrm{OB}}^+ / \mathbb{R} \cong S^1.$$

The curves in $\mathcal{M}_{\mathrm{OB}}^+$ have the "correct" index, in the sense that the actual and virtual dimensions match. In [Wen10c], it is shown, in fact, that for suitable (nongeneric!) choices of data, an open book with pages of *any* genus $g \geq 0$ admits a 2-parameter family of pseudoholomorphic lifts, but they have index $2 - 2g$, which is the correct virtual dimension only when $g = 0$. This is why Theorem 5.6 fails in general for open books of positive genus (cf. Remark A.9).

Now suppose (W, ω) is a symplectic filling of (M, ξ), where the latter is supported by a planar open book. By modifying ω near ∂W, we can assume without loss of generality (possibly after rescaling ω) that it takes the form $d(e^s \alpha)$ in a collar neighborhood of the boundary, where α is the contact form provided by Theorem 5.10. Let $(\widehat{W}, \widehat{\omega})$ denote the resulting symplectic

5.3 Sketch of the Proof

completion, and choose an $\hat{\omega}$-compatible almost complex structure J that is generic in W and matches J_+ (from Theorem 5.10) on $[0, \infty) \times M$. Since the J_+-holomorphic pages in $\mathbb{R} \times M$ have no negative punctures, each can be assumed to lie in $[0, \infty) \times M$ after some $\mathbb{R}$-translation, so these give rise to a 2-dimensional family of embedded J-holomorphic curves living in the cylindrical end of $\hat{W}$, which we shall refer to henceforward as the *J-holomorphic pages in $\hat{W}$*. Let

$$\mathcal{M}_{\mathrm{OB}} \subset \mathcal{M}_0(\hat{W}, J)$$

denote the connected component of the moduli space $\mathcal{M}_0(\hat{W}, J)$ that contains these J-holomorphic pages, and let $\overline{\mathcal{M}}_{\mathrm{OB}}$ denote its closure in the compactified moduli space $\overline{\mathcal{M}}_0(\hat{W}, J)$ (see Appendix A.2). Theorem 5.6 can be deduced from the following:

Proposition 5.11 *The compactified moduli space $\overline{\mathcal{M}}_{\mathrm{OB}}$ is diffeomorphic to the 2-disk, and the elements of $\overline{\mathcal{M}}_{\mathrm{OB}}$ can be described as follows:*

- *The smooth curves in $\mathcal{M}_{\mathrm{OB}}$ are all embedded and pairwise disjoint, and they foliate $\hat{W}$ outside a finite union of properly embedded surfaces.*
- *There is a natural identification*

$$\partial \overline{\mathcal{M}}_{\mathrm{OB}} = \mathcal{M}_{\mathrm{OB}}^+/\mathbb{R},$$

  where each $\mathbb{R}$-equivalence class of J_+-holomorphic pages $u_P \in \mathcal{M}_{\mathrm{OB}}^+$ in $\mathbb{R} \times M$ is identified with a holomorphic building that has empty main level and a single upper level consisting of u_P.[2]
- *There are at most finitely many elements of $\overline{\mathcal{M}}_{\mathrm{OB}} \backslash \mathcal{M}_{\mathrm{OB}}$ in the interior of $\overline{\mathcal{M}}_{\mathrm{OB}}$, and they are pairwise disjoint nodal curves in $\hat{W}$, each having exactly two connected components, which are embedded and intersect each other exactly once, transversely. Any such component that is closed also has homological self-intersection number -1.*

Every point in $\hat{W}$ lies in the image of a unique (possibly nodal) curve in the interior of $\overline{\mathcal{M}}_{\mathrm{OB}}$.

To prove this, notice first that all curves in $\mathcal{M}_{\mathrm{OB}}$ are guaranteed to be simple, since their asymptotic orbits are all distinct and simply covered. By Theorem 4.17, all $u \in \mathcal{M}_{\mathrm{OB}}$ then satisfy $\delta_\infty(u) = 0$, as double points can only be hidden at infinity if there exist multiply covered asymptotic orbits or two

---

[2] It is standard to define the space of holomorphic buildings $\overline{\mathcal{M}}_g(\hat{W}, J)$ such that two buildings are considered equivalent if they differ only by an $\mathbb{R}$-translation of one of the symplectization levels (see [BEH+03]).

88   *Symplectic Fillings of Planar Contact 3-Manifolds*

distinct punctures asymptotic to coinciding orbits. Since $\delta(u) + \delta_\infty(u)$ is homotopy invariant (Theorem 4.4), and the J-holomorphic pages u_P are embedded and thus satisfy $\delta(u_P) = 0$, we conclude

$$\delta(u) = \delta_\infty(u) = 0 \quad \text{for all } u \in \mathcal{M}_{\text{OB}};$$

hence all curves in $\mathcal{M}_{\text{OB}}$ are embedded. Theorem 3.26 then implies slightly more: since the curves in $\mathcal{M}_{\text{OB}}$ also have index 2 and genus 0 and all their asymptotic orbits have odd Conley–Zehnder index, we have:

Lemma 5.12 *For each $u \in \mathcal{M}_{\text{OB}}$, there is a neighborhood $\mathcal{U} \subset \mathcal{M}_{\text{OB}}$ such that the curves in $\mathcal{U}$ are all embedded and their images foliate a neighborhood of the image of u in $\widehat{W}$.*

The self-intersection number $u * u$ for any curve $u \in \mathcal{M}_{\text{OB}}$ can now be computed easily from Siefring's adjunction formula (4.1): since all asymptotic orbits are simply covered, the spectral covering term $\bar{\sigma}(u) - \#\Gamma$ vanishes, and so does $c_N(u)$ due to formula (3.18); thus

$$u * u = 2\left[\delta(u) + \delta_\infty(u)\right] + c_N(u) + \left[\bar{\sigma}(u) - \#\Gamma\right] = 0.$$

This result can alternatively be deduced as an immediate corollary of Lemma 5.15 below. By Theorem 4.1, we conclude:

Lemma 5.13 *Any two distinct curves in $\mathcal{M}_{\text{OB}}$ are disjoint.*

Combining that with Lemma 5.12, it follows that the curves in $\mathcal{M}_{\text{OB}}$ form a smooth foliation of some open subset of $\widehat{W}$.

Lemma 5.14 *Assume $u_P \in \mathcal{M}^+_{\text{OB}}$ is a J_+-holomorphic page in $\mathbb{R} \times M$, and $u_\gamma \colon \mathbb{R} \times S^1 \to \mathbb{R} \times M$ is the orbit cylinder for an embedded Reeb orbit γ that is a component of the binding B. Then*

$$u_P * u_\gamma = 0.$$

Proof The page $P \subset M$ is always disjoint from the binding $B \subset M$; thus $u_P \cdot u_\gamma = 0$, so it only remains to show that $\iota_\infty(u_P, u_\gamma) = 0$. By Theorem 4.13, this is true if and only if the asymptotic eigenfunction controlling the approach of the relevant end of u_P to γ has extremal winding. Let τ denote the trivialization of $\gamma^*\xi$ in which pages approach γ with winding number 0; hence, by construction, the winding (relative to τ) of the eigenfunction in question is 0. Combining Theorem 5.10 with Proposition 3.17, we also have

$$1 = \mu^\tau_{\text{CZ}}(\gamma) = 2\alpha^\tau_-(\gamma) + p(\gamma);$$

hence the extremal winding is also $\alpha^\tau_-(\gamma) = 0$, and this implies $\iota_\infty(u_P, u_\gamma) = 0$ as claimed. $\qquad\qquad\square$

5.3 Sketch of the Proof

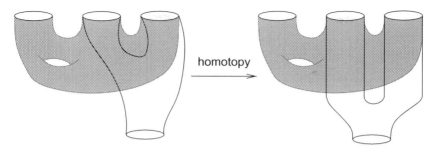

Figure 5.4 A homotopy through asymptotically cylindrical maps for the proof of Lemma 5.15.

Lemma 5.15 *Assume $u_P \in \mathcal{M}_{OB}^+$ is a J_+-holomorphic page in $\mathbb{R} \times M$ and $v: \dot{\Sigma}' \to \mathbb{R} \times M$ is any J_+-holomorphic curve whose positive ends are all asymptotic to embedded Reeb orbits in the binding B. Then $u_P * v = 0$.*

Proof The argument depends only on the following facts:

(1) u_P has no negative ends.
(2) By Lemma 5.14, $u_P * u_\gamma = 0$ for all orbits γ that appear at positive ends of v, where u_γ denotes the orbit cylinder over γ.

Figure 5.4 shows a homotopy through asymptotically cylindrical maps for two curves satisfying the above conditions (u_P has positive genus in the picture, which has no impact on the argument). After a homotopy, we may assume namely that u_P lives entirely in $[0, \infty) \times M$, while the portion of v living in $[0, \infty) \times M$ is simply a disjoint union of orbit cylinders u_γ for which Lemma 5.14 implies $u * u_\gamma = 0$. Using the homotopy invariance[3] of the $*$-pairing, we conclude that $u * v$ equals a sum of terms of the form $u * u_\gamma$, all of which vanish. □

Lemma 5.16 (cf. [Sie11, Theorem 5.21]) *Other than the J_+-holomorphic pages $u_P \in \mathcal{M}_{OB}^+$ and the orbit cylinders over embedded orbits in B, there exist no J_+-holomorphic curves in $\mathbb{R} \times M$ that are asymptotic to embedded orbits in B at all their positive punctures.*

Proof Any such curve $v: \dot{\Sigma}' \to \mathbb{R} \times M$ must intersect one of the pages u_P, as these foliate $\mathbb{R} \times (M \backslash B)$; hence

$$u_P * v \geq u_P \cdot v > 0,$$

and this contradicts Lemma 5.15. □

---

[3] Note that the homotopy in our proof of Lemma 5.15 is *not* a homotopy through J-holomorphic curves, but only through asymptotically cylindrical maps.

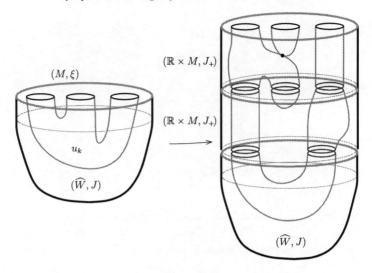

Figure 5.5 A hypothetical degeneration of a sequence $u_k \in \mathcal{M}_{\text{OB}}$ to a holomorphic building in $\overline{\mathcal{M}}_{\text{OB}}$. This scenario is ruled out by Lemma 5.16, which says that the two curves in the top level that are not orbit cylinders cannot exist.

We are now in a position to justify the description of the compactification $\overline{\mathcal{M}}_{\text{OB}}$ given in Proposition 5.11.

Lemma 5.17 *Suppose $u_k \in \mathcal{M}_{\text{OB}}$ is a sequence convergent to a holomorphic building with at least one nontrivial upper level. Then the main level of the limit is empty, and its upper level is a J_+-holomorphic page in $\mathcal{M}_{\text{OB}}^+$.*

Proof If the lemma is false, then we obtain a holomorphic building whose top level contains a J_+-holomorphic curve in $\mathbb{R} \times M$ that is not in $\mathcal{M}_{\text{OB}}^+$ but has all its positive punctures asymptotic to orbits in the binding B (see Figure 5.5). This is impossible by Lemma 5.16. □

We can now identify $\partial \overline{\mathcal{M}}_{\text{OB}}$ with $\mathcal{M}_{\text{OB}}^+/\mathbb{R} \cong S^1$ as described in Proposition 5.11, and the above lemma says that all other elements of $\overline{\mathcal{M}}_{\text{OB}} \backslash \mathcal{M}_{\text{OB}}$ must be buildings with no upper levels, i.e., nodal J-holomorphic curves in $\widehat{W}$. The components of these nodal curves have only positive ends (if any), all asymptotic to distinct simply covered orbits in the binding B. These orbits all have have Conley–Zehnder index 1 relative to the canonical trivialization. Now (A.5) gives the index of such a curve $v \colon \Sigma' \backslash \Gamma' \to \widehat{W}$ as

$$\text{ind}(v) = -\chi(\Sigma' \backslash \Gamma') + 2c_1^\tau(v^*T\widehat{W}) + \sum_{z \in \Gamma'} \mu_{\text{CZ}}^\tau(\gamma_z) = -\chi(\Sigma') + 2c_1^\tau(v^*T\widehat{W}).$$

5.3 Sketch of the Proof

This matches the index formula for the closed case closely enough that one can now repeat the compactness argument in the proof of Lemma 1.17 (see Appendix A.1) more or less verbatim,[4] thus proving the following:

Lemma 5.18 *There exists a finite set of simple curves* $\mathcal{B} \subset M_0(\widehat{W}, J)$, *with index* 0, *such that every nodal curve in* $\overline{\mathcal{M}}_{\mathrm{OB}}$ *has exactly two components* $v_+, v_- \in \mathcal{B}$. $\qquad\square$

To finish the proof, we must study the intersection-theoretic properties of the components of nodal curves $\{v_+, v_-\} \in \overline{\mathcal{M}}_{\mathrm{OB}}$. Given such a curve as limit of a sequence $u_k \in \mathcal{M}_{\mathrm{OB}}$, we have

$$0 = u_k * u_k = v_+ * v_+ + v_- * v_- + 2(v_+ * v_-). \tag{5.1}$$

Exercise 5.19 Verify (5.1), using the definition of the $*$-pairing from Lecture 4.

Observe that v_+ and v_- cannot be the same curve up to parametrization, as they are required to have distinct sets of asymptotic orbits. This implies that they have at least one isolated intersection, so by Theorem 4.1,

$$v_+ * v_- \geq v_+ \cdot v_- \geq 1. \tag{5.2}$$

Since $\mathrm{ind}(v_\pm) = 0$ and all asymptotic orbits of $v_\pm$ have odd Conley–Zehnder index, (3.18) gives

$$c_N(v_\pm) = -1.$$

Now applying Siefring's adjunction formula (Theorem 4.4), the spectral covering numbers $\bar{\sigma}(v_\pm)$ are each equal to the number of punctures since all orbits are simply covered, so these terms vanish from the adjunction formula, and we have

$$v_\pm * v_\pm = 2\left[\delta(v_\pm) + \delta_\infty(v_\pm)\right] + c_N(v_\pm) = 2\left[\delta(v_\pm) + \delta_\infty(v_\pm)\right] - 1.$$

Combining this with (5.1) gives

$$0 = 2 \sum_\pm \left[\delta(v_\pm) + \delta_\infty(v_\pm)\right] + 2(v_+ * v_- - 1),$$

so in light of (5.2), we have

$$\delta(v_\pm) = \delta_\infty(v_\pm) = 0, \qquad v_+ * v_- = v_+ \cdot v_- = 1, \qquad \text{and} \qquad v_\pm * v_\pm = -1,$$

---

[4] There are two main differences from the closed case: first, it is trivial to prove that non-closed curves in $\mathcal{B}$ are simple, since their asymptotic orbits are distinct and simply covered, while for closed components one must apply the same argument as before. Second, proving compactness (and hence finiteness) of $\mathcal{B}$ requires first ruling out holomorphic buildings with nontrivial upper levels – this works the same way as in Lemma 5.17.

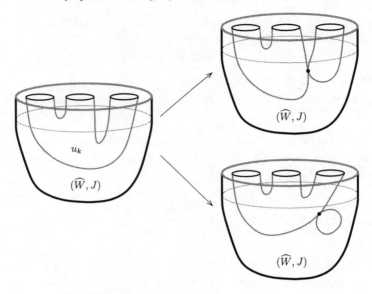

Figure 5.6 Two possible degenerations of a sequence $u_k \in \mathcal{M}_{OB}$ to nodal curves in $\overline{\mathcal{M}}_{OB}$. The second scenario includes a J-holomorphic exceptional sphere and is thus only possible if (W, ω) is not minimal.

implying that $v_\pm$ are both embedded and intersect each other exactly once, transversely. Moreover, if either component is closed, then its homological self-intersection number is now $v_\pm \cdot v_\pm = v_\pm * v_\pm = -1$; hence it is a J-holomorphic exceptional sphere (see Figure 5.6). In the same manner, one can show that the nodal curves in $\overline{\mathcal{M}}_{OB}$ are all fully disjoint from each other and from the smooth curves in $\mathcal{M}_{OB}$.

Arguing as in the proof of Theorem 1.16, let $F \subset \widehat{W}$ denote the union of the images of the curves in $\mathcal{B}$, which are finitely many properly embedded surfaces. Then let

$$X := \left\{ p \in \widehat{W} \backslash F \,\middle|\, p \text{ is in the image of a curve in } \mathcal{M}_{OB} \right\}.$$

The lemmas proved in the preceding pages imply that X is an open and closed subset of $\widehat{W} \backslash F$; thus $X = \widehat{W} \backslash F$, and we see that every point in $\widehat{W}$ is in the image of a unique (possibly nodal) curve in $\overline{\mathcal{M}}_{OB}$, giving a surjective map

$$\Pi \colon \widehat{W} \to \overline{\mathcal{M}}_{OB} \backslash \partial \overline{\mathcal{M}}_{OB}.$$

Since the J_+-holomorphic pages $[u_P] \in \mathcal{M}_{OB}^+ / \mathbb{R} = \partial \overline{\mathcal{M}}_{OB}$ also foliate $M \backslash B$ under the projection $\mathbb{R} \times M \to M$, we can extend Π to the natural compactification $\overline{W} := \widehat{W} \cup (\{\infty\} \times M)$ as a surjective map

5.3 Sketch of the Proof

$$\Pi \colon \overline{W} \backslash B \to \overline{\mathcal{M}}_{\mathrm{OB}},$$

whose smooth fibers are the compact symplectically embedded surfaces with boundary obtained as the images of maps $\bar{u} \colon \overline{\Sigma} \to \overline{W}$ for $u \in \mathcal{M}_{\mathrm{OB}}$, and we are treating B as a submanifold of $\{\infty\} \times M = \partial \overline{W}$. There is still a small amount of work to be done in identifying the above construction with something that one can regard as a smooth symplectic Lefschetz fibration; details (in a more general setting) may be found in [LVWb].

One detail in Proposition 5.11 has not yet been verified: we've seen that $\mathcal{M}_{\mathrm{OB}}$ is an oriented 2-dimensional manifold, compactified by adding finitely many interior points (the nodal curves) and the boundary $\partial \overline{\mathcal{M}}_{\mathrm{OB}} = \mathcal{M}_{\mathrm{OB}}^{+}/\mathbb{R} \cong S^1$; hence $\overline{\mathcal{M}}_{\mathrm{OB}}$ is a compact oriented surface with one boundary component, but we claim, in fact, that it is a *disk*. To see this, choose a smooth loop $\gamma \colon S^1 \to M$ near a binding component that meets every page of the open book exactly once transversely. Viewing γ as a loop in $\{\infty\} \times M = \partial \overline{W}$, the loop

$$\Pi \circ \gamma \colon S^1 \to \overline{\mathcal{M}}_{\mathrm{OB}}$$

then parametrizes $\partial \overline{\mathcal{M}}_{\mathrm{OB}}$. Now, γ is obviously not contractible in $M \backslash B$, but we can easily assume it is contractible in $\overline{W} \backslash B$: indeed, γ can be chosen contractible in M, and then translating downward from $\{\infty\} \times M$ to a level $\{s\} \times M \subset \widehat{W}$ for $s \in [0, \infty)$ gives a contractible loop in $\widehat{W}$. Composing this contraction with Π, we conclude

$$[\partial \overline{\mathcal{M}}_{\mathrm{OB}}] = 0 \in \pi_1(\overline{\mathcal{M}}_{\mathrm{OB}});$$

hence $\overline{\mathcal{M}}_{\mathrm{OB}} \cong \mathbb{D}$.

Appendix A

Properties of Pseudoholomorphic Curves

In this appendix we will summarize (without proofs) the essential global analytical results about pseudoholomorphic curves that are used in various places in these lectures. The first section covers results on closed holomorphic curves that are needed in Lectures 1 and 2, and the second section then states the generalizations of these results to punctured curves in completed symplectic cobordisms. For more details on each, we refer to [MS12] or [Wena] for the closed case and [Wenb] for the punctured case.

A.1 The Closed Case

Given a closed symplectic manifold (M, ω) with a compatible[1] almost complex structure J, we defined in §1.2 the moduli space $\mathcal{M}_g^A(M, J)$ of (equivalence classes up to parametrization of) J-holomorphic curves with genus $g \geq 0$ homologous to $A \in H_2(M)$. We shall now summarize the main analytical properties of this space and use them to prove Lemma 1.17.

The **virtual dimension** of $\mathcal{M}_g^A(M, J)$, also sometimes called the **index** of a curve $u \in \mathcal{M}_g^A(M, J)$ and denoted by $\mathrm{ind}(u) \in \mathbb{Z}$, is defined to be the integer

$$\text{vir-dim}\, \mathcal{M}_g^A(M, J) := (n - 3)(2 - 2g) + 2c_1(A), \tag{A.1}$$

where $c_1(A)$ is shorthand for the evaluation of the first Chern class $c_1(TM, J) \in H^2(M)$ on the homology class A. This definition of vir-dim $\mathcal{M}_g^A(M, J)$ is justified by Theorem A.3.

---

[1] The vast majority of the results we will state here can also be generalized for almost complex structures that are *tamed* by ω but not necessarily compatible. Such generalizations become much less straightforward whenever asymptotic analysis is involved; thus, in the punctured case, it is best always to assume J is compatible and not just tame.

A.1 The Closed Case

Recall that closed J-holomorphic curves $u\colon (\Sigma, j) \to (M, J)$ are always either **simple** or **multiply covered**, where the latter means $u = v \circ \varphi$ for some closed J-holomorphic curve $v\colon (\Sigma', j') \to (M, J)$ and holomorphic map $\varphi\colon (\Sigma, j) \to (\Sigma', j')$ of degree $\deg(\varphi) > 1$. By a slight abuse of terminology (see Remark 2.5), we refer to a point $z \in \Sigma$ in the domain of a J-holomorphic curve $u\colon \Sigma \to M$ as a **critical point** if $du(z) = 0$; the alternative is that $du(z)\colon T_z\Sigma \to T_{u(z)}M$ is injective, in which case we call z an **immersed point**. Combining general topological arguments with the local properties of J-holomorphic curves (e.g., Theorem B.23 in Appendix B), one can show the following:

Theorem A.1 *Every nonconstant, closed and connected J-holomorphic curve $u\colon (\Sigma, j) \to (M, J)$ has at most finitely many critical points. Moreover, if u is simple, then it also has at most finitely many double points; hence it is embedded outside of some finite subset of Σ.*

The following related result is sometimes referred to as the **unique continuation** principle:

Theorem A.2 *If u and v are two closed J-holomorphic curves that are both simple, then they are either equivalent up to parametrization or have at most finitely many intersections.*

The **automorphism group** of a triple (Σ, j, u) representing an element of $\mathcal{M}_g^A(M, J)$ is defined as

$$\mathrm{Aut}\,(\Sigma, j, u) = \{\varphi\colon (\Sigma, j) \to (\Sigma, j) \text{ biholomorphic} \mid u = u \circ \varphi\}.$$

This group is always finite if $u\colon \Sigma \to M$ is not constant, and Theorem A.1 implies that it is trivial whenever u is simple.

The following result is dependent on a definition of the term **Fredholm regular**, which is rather technical; therefore, we will not give it – this is obviously a terrible thing to do, but we hope Theorems A.4 and A.6 will make up for it. The proofs of these results depend on the regularity theory of elliptic PDEs (see [MS12] or [Wena] for details).

Theorem A.3 *The subset of $\mathcal{M}_g^A(M, J)$ consisting of all curves that are Fredholm regular and have trivial automorphism groups is* open, *and, moreover, it naturally admits the structure of a smooth oriented finite-dimensional manifold, with dimension equal to* vir-dim $\mathcal{M}_g^A(M, J)$.

96 Properties of Pseudoholomorphic Curves

Recall that for any topological space X, a subset $Y \subset X$ is said to be **comeager** if it contains a countable intersection of open dense sets.[2] If X is a complete metric space, then the Baire category theorem implies that every comeager subset of X is also dense.

Theorem A.4 *Suppose (M, ω) is a closed symplectic manifold, $\mathcal{U} \subset M$ is an open subset, and J_0 is an ω-compatible almost complex structure on M. Let $\mathcal{J}(\mathcal{U}, J_0)$ denote the space of all smooth ω-compatible almost complex structures J on M such that $J \equiv J_0$ on $M \backslash \mathcal{U}$, and assign to $\mathcal{J}(\mathcal{U}, J_0)$ the natural C^∞-topology. Then there exists comeager subset $\mathcal{J}^{\mathrm{reg}}(\mathcal{U}, J_0) \subset \mathcal{J}(\mathcal{U}, J_0)$ such that, for all $J \in \mathcal{J}^{\mathrm{reg}}(\mathcal{U}, J_0)$, every simple curve $u \in \mathcal{M}_g^A(M, J)$ that intersects $\mathcal{U}$ is Fredholm regular.*

Results such as Theorem A.4 that hold for all data in some comeager subset are often said to hold for **generic** data, so one can summarize the two theorems above by saying that the moduli space of simple J-holomorphic curves is a smooth manifold of the "correct" dimension for "generic" J. This fact is true even for moduli spaces with vir-dim $\mathcal{M}_g^A(M, J) < 0$, implying that, in such spaces, no Fredholm regular curves exist.

Corollary A.5 *For generic ω-compatible almost complex structures J in a closed symplectic manifold (M, ω), every simple J-holomorphic curve u satisfies $\mathrm{ind}(u) \geq 0$.*

Theorem A.4 is a "transversality" result; i.e., it follows from an infinite-dimensional version (the *Sard-Smale theorem*) of the standard fact from differential topology that any two submanifolds intersect each other transversely after a generic perturbation. Occasionally, one also needs transversality results for nongeneric data. Such results exist – they follow from the Riemann–Roch formula in certain fortunate situations – but their utility is typically limited to dimension 4 and genus 0. The following theorem of Hofer, Lizan and Sikorav [HLS97] is closely related to the question of local foliations discussed in §1.3.

Theorem A.6 *If $\dim M = 4$ and J is any almost complex structure on M, then every immersed J-holomorphic curve $u \in \mathcal{M}_g^A(M, J)$ with $\mathrm{ind}\,(u) > 2g - 2$ is Fredholm regular.*

---

[2] It is common among symplectic topologists to say that comeager subsets are "Baire sets" or are "of second category," but this seems to be slightly inconsistent with the standard usage of these terms in other fields.

A.1 The Closed Case

The moduli space $\mathcal{M}_g^A(M,J)$ is not generally compact, but if M is closed and J is compatible with a symplectic form ω, then it has a natural *compactification*. The **energy** of a curve $u \in \mathcal{M}_g(M,J)$ can be defined as

$$E(u) = \int_\Sigma u^*\omega$$

for any parametrization $u\colon \Sigma \to M$; the taming condition implies that $E(u) \geq 0$ for all J-holomorphic curves, with equality if and only if the curve is constant. Observe that $E(u)$ only depends on $[u] \in H_2(M)$.

The moduli space of **stable nodal J-holomorphic curves of arithmetic genus** g homologous to $A \in H_2(M)$ is defined as

$$\overline{\mathcal{M}}_g^A(M,J) := \{(S,j,u,\Delta)\} \big/ \sim,$$

where

- (S,j) is a (possibly disconnected) closed Riemann surface.
- The set of **nodes**, $\Delta \subset S$, is a finite unordered set of pairwise distinct points organized into pairs

$$\Delta = \{\{\hat{z}_1, \check{z}_1\}, \dots, \{\hat{z}_r, \check{z}_r\}\}$$

  such that the singular surface

$$\hat{S} := S\big/\hat{z}_j \sim \check{z}_j \text{ for } j = 1, \dots, r$$

  is homeomorphic to a (possibly singular) fiber of some Lefschetz fibration with regular fibers of genus g.
- $u\colon (S,j) \to (M,J)$ is a pseudoholomorphic map with $[u] = A$ that descends to the quotient $\hat{S} = S/\sim$ as a continuous map $\hat{S} \to M$.
- Every connected component of $S\backslash\Delta$ on which u is constant has negative Euler characteristic.
- We write $(S,j,u,\Delta) \sim (S',j',u',\Delta')$ if there is a biholomorphic map $\varphi\colon (S,j) \to (S',j')$ such that $u = u' \circ \varphi$ and φ maps pairs in Δ to pairs in Δ'.

The condition on the Euler characteristics of constant components is called **stability** – its effect is to exclude certain ambiguities that would otherwise cause nonuniqueness of limits for the natural topology on $\overline{\mathcal{M}}_g^A(M,J)$. Assuming $A \neq 0$ so that elements of $\mathcal{M}_g^A(M,J)$ are never constant, there is a natural inclusion $\mathcal{M}_g^A(M,J) \subset \overline{\mathcal{M}}_g^A(M,J)$ defined by setting $\Delta := \varnothing$ for any $[(\Sigma,j,u)] \in \mathcal{M}_g^A(M,J)$. We denote the union over all $A \in H_2(M)$ by $\overline{\mathcal{M}}_g(M,J)$.

98 Properties of Pseudoholomorphic Curves

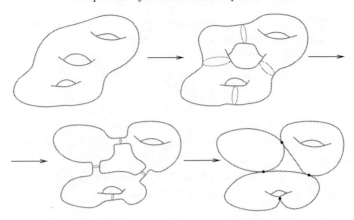

Figure A.1 A sequence of genus 3 holomorphic curves degenerating to a nodal curve of arithmetic genus 3, with four nodes, two connected components of genus 0 and one of genus 1.

Theorem A.7 (*Gromov's compactness theorem*) *For each $A \in H_2(M)$, $g \geq 0$ and each ω-compatible almost complex structure J on a closed symplectic manifold (M, ω), $\overline{\mathcal{M}}_g^A(M, J)$ admits a natural topology as a compact metrizable space. Moreover, any sequence $u_k \in \mathcal{M}_g(M, J)$ of curves satisfying a uniform energy bound $E(u_k) \leq C$ has a subsequence convergent to an element of $\overline{\mathcal{M}}_g(M, J)$.*

Remark A.8 The second statement in the above theorem does not impose any direct restriction on the homology classes $[u_k] \in H_2(M)$, but it *implies* the existence of a subsequence with constant homology. Observe that the required energy bound is automatic if all u_k represent a fixed homology class.

It will not be necessary for our purposes to give a complete definition of the topology of $\overline{\mathcal{M}}_g^A(M, J)$, but we can describe the convergence of a sequence of smooth curves $[(\Sigma_k, j_k, u_k)] \in \mathcal{M}_g^A(M, J)$ to a nodal curve $[(S, j, u, \Delta)] \in \overline{\mathcal{M}}_g^A(M, J)$ as follows (see Figure A.1 for an example): let S' denote the compact topological 2-manifold with boundary (Figure A.2, lower left) obtained from S by replacing each point $z \in \Delta \subset S$ with the circle

$$C_z := T_z S / \mathbb{R}_+,$$

where $\mathbb{R}_+ := (0, \infty)$ acts on $T_z S$ by scalar multiplication. The smooth structure of $S \backslash \Delta$ does not have an obviously canonical extension over S', but each boundary component $C_z \subset \partial S'$ inherits from the conformal structure of (S, j) a natural class of preferred diffeomorphisms to S^1. Now since the points in

A.1 The Closed Case

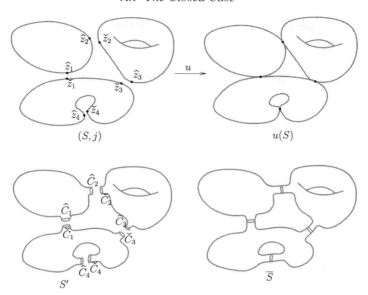

Figure A.2 Four ways of viewing the nodal holomorphic of Figure A.1. At the upper left, we see the disconnected Riemann surface (S, j) with nodal pairs $\{\hat{z}_i, \check{z}_i\}$ for $i = 1, 2, 3, 4$. To the right of this is a possible picture of the image of the nodal curve, with nodal pairs always mapped to identical points. The bottom right shows the surface S' with boundary, obtained from S by replacing the points $\hat{z}_i, \check{z}_i$ with circles $\hat{C}_i, \check{C}_i$. Gluing these pairs of circles together gives the closed connected surface $\overline{S}$ at the bottom right, whose genus is by definition the arithmetic genus of the nodal curve.

Δ come in pairs $\{\hat{z}, \check{z}\}$, we can make a choice of preferred orientation-reversing diffeomorphisms $C_{\hat{z}} \to C_{\check{z}}$ for each such pair and glue corresponding boundary components of S' to define a closed surface (Figure A.2, lower right),

$$\overline{S} := S'/C_{\hat{z}} \sim C_{\check{z}}.$$

This is naturally a closed topological 2-manifold, and it also carries a smooth structure and a complex structure on $\overline{S} \setminus C = S \setminus \Delta$, where

$$C := \bigcup_{z \in \Delta} C_z \subset \overline{S}.$$

Since $u(\hat{z}) = u(\check{z})$ for each pair $\{\hat{z}, \check{z}\} \subset \Delta$, u extends from $\overline{S} \setminus C$ to a continuous map

$$\bar{u} : \overline{S} \to M.$$

Properties of Pseudoholomorphic Curves

The convergence $[(\Sigma_k, j_k, u_k)] \to [(S, j, u, \Delta)]$ can now be defined to mean that for sufficiently large k there exist homeomorphisms

$$\varphi_k \colon \overline{S} \to \Sigma_k$$

whose restrictions to $\overline{S} \backslash C$ are smooth and have smooth inverses such that

$$\varphi_k^* j_k \to j \text{ in } C_{\mathrm{loc}}^\infty(\overline{S} \backslash C), \quad u_k \circ \varphi_k \to u \text{ in } C_{\mathrm{loc}}^\infty(\overline{S} \backslash C, M), \quad \text{and}$$

$$u_k \circ \varphi_k \to \bar{u} \text{ in } C^0(\overline{S}, M).$$

The analytical toolbox is now complete enough to fill in the following gap from §1.2.

Proof of Lemma 1.17 By construction, $\mathcal{M}_0^{[S]}(M, J)$ contains an embedded curve u_S, defined as the inclusion of S. The almost complex structure J cannot be assumed "generic" in the sense of Theorem A.4 since we chose it specifically to have the property of preserving TS. We claim, however, that u_S is nonetheless Fredholm regular due to Theorem A.6. Indeed, it has trivial normal bundle $N_S \to S^2$ since $[S] \cdot [S] = 0$, so the natural splitting of complex vector bundles

$$(u_S^* TM, J) = (TS^2, j) \oplus (N_S, J)$$

implies

$$c_1([S]) := c_1(u_S^* TM) = c_1(TS^2) + c_1(N_S) = \chi(S^2) + 0 = 2.$$

Plugging $c_1([S]) = 2$ and $n = 2$ into the index formula (A.1) now gives

$$\mathrm{ind}(u_S) = -2 + 2c_1([S]) = 2;$$

hence vir-dim $\mathcal{M}_0^{[S]}(M, J) = 2$. Since u_S also is immersed, it now satisfies the hypotheses of Theorem A.6, so Fredholm regularity follows.

To achieve smoothness near the rest of the simple curves in $\mathcal{M}_0^{[S]}(M, J)$, it suffices to choose a generic perturbation J' of J on the open subset $M \backslash S$. Indeed, for any such J', assuming $J' = J$ along S ensures that u_S is also J'-holomorphic, so unique continuation (Theorem A.2) then implies that no other J'-holomorphic curve in M can be contained entirely in S unless it is a multiple cover of u_S. In particular, u_S itself is the only such curve that is either simple or homologous to $[S]$. It follows, then, by Theorem A.4, that every other simple curve in $\mathcal{M}_0(M, J')$ is also Fredholm regular, so, by Theorem A.3, the subset $\mathcal{M}_0^{[S],*}(M, J') \subset \mathcal{M}_0^{[S]}(M, J')$ of simple curves is an oriented 2-dimensional manifold. To simplify the notation, we relabel $J := J'$ from now on.

By Gromov's compactness theorem (Theorem A.7), any sequence $u_k \in \mathcal{M}_0^{[S],*}(M, J)$ with no convergent subsequence in $\mathcal{M}_0^{[S]}(M, J)$ converges to a

A.1 The Closed Case

nodal curve with arithmetic genus 0. The genus condition implies that its connected components are all spheres, so we can regard the nodal curve simply as a finite set of J-holomorphic spheres $v_1, \ldots, v_N \in \mathcal{M}_0(M, J)$ with $N \geq 2$, satisfying the condition

$$[v_1] + \cdots + [v_N] = [S]. \tag{A.2}$$

These spheres cannot at first be assumed to be simple, but for each $j = 1, \ldots, N$, there is a simple curve $w_j \in \mathcal{M}_{g_j}(M, J)$ and an integer $k_j \in \mathbb{N}$ such that v_j is a k_j-fold cover of w_j; here we adopt the convention $w_j = v_j$ if $k_j = 1$. If $k_j > 1$, then v_j factors through a holomorphic map $S^2 \to \Sigma_{g_j}$ of degree k_j, where Σ_{g_j} is a closed connected surface with genus g_j; but no such map exists if $g_j > 0$ since the universal cover of Σ_{g_j} is then contractible, implying $\pi_2(\Sigma_{g_j}) = 0$, so we conclude that each w_j has genus 0. Now since all simple J-holomorphic curves in M are Fredholm regular, Corollary A.5 and the index formula (A.1) give

$$\mathrm{ind}(w_j) = -2 + 2c_1([w_j]) \geq 0;$$

hence $c_1([w_j]) \geq 1$. Since $c_1([S]) = 2$, (A.2) now gives

$$k_1 + \cdots + k_N \leq k_1 c_1([w_1]) + \cdots + k_N c_1([w_N]) = 2; \tag{A.3}$$

thus $N = 2$ and $k_1 = k_2 = c_1([v_1]) = c_1([v_2]) = 1$. We conclude that the nodal curve has exactly two components, both simple, and, since $[v_1] + [v_2] = [S]$, they satisfy the uniform energy bound

$$E(v_j) = \langle [\omega], [v_j] \rangle \leq \langle [\omega], [v_1] \rangle + \langle [\omega], [v_2] \rangle = \langle [\omega], [S] \rangle \tag{A.4}$$

for $j = 1, 2$.

Finally, we claim that the set of simple curves $v \in \mathcal{M}_0(M, J)$ with $c_1([v]) = 1$ is finite. By Theorems A.3 and A.4, this set is a 0-dimensional manifold, i.e., a discrete set, so finiteness will follow if we can show that it is compact. This follows essentially by a repeat of the argument above; note that Gromov's compactness theorem is applicable due to the energy bound (A.4). Now if a sequence of such curves converges to a nodal curve with more than one component, then it produces an inequality like (A.3) but with 1 on the right-hand side, which gives a contradiction. The only remaining possibility is that a sequence v_k of curves with $c_1([v_k]) = 1$ converges to a smooth but multiply covered curve v, but this is immediately excluded since $c_1([v]) = 1$, so $[v]$ is a primitive homology class. $\qquad\square$

Remark A.9 Let us see what goes wrong if one tries to prove an analogue of McDuff's theorem about ruled surfaces under the assumption of a symplectically embedded surface $S \subset (M, \omega)$ with $[S] \cdot [S] = 0$ and genus $g > 0$. One can

102   *Properties of Pseudoholomorphic Curves*

still construct an embedded J-holomorphic curve $u_S \in \mathcal{M}_g^{[S]}(M, J)$, and since its normal bundle $N_S \to S$ is necessarily trivial, the splitting $u^*TM = TS \oplus N_S$ now gives $c_1([S]) = \chi(S) = 2 - 2g$, so (A.1) now gives

$$\text{vir-dim } \mathcal{M}_g^{[S]}(M, J) = -(2 - 2g) + 2c_1([S]) = 2 - 2g.$$

This answer is desirable when $g = 0$ because 2 is the right number of dimensions to foliate a 4-manifold by holomorphic curves – but if $g > 0$, one cannot hope to find a 2-parameter family of holomorphic curves homologous to $[S]$, and, in fact, the curves should disappear entirely after a generic perturbation if $g > 1$. The failure of the proof is thus attributable essentially to the Riemann–Roch formula, from which the dimension formula (A.1) is derived. It is more than a failure of technology, however, as the theorem is false when $g > 0$.

A.2 Curves with Punctures

A general reference for the contents of this section is [Wenb].

Assume (W, ω) is a $2n$-dimensional symplectic cobordism with

$$\partial(W, \omega) = (-M_-, \xi_- = \ker \alpha_-) \sqcup (M_+, \xi_+ = \ker \alpha_+),$$

$(\widehat{W}, \widehat{\omega})$ denotes its completion and $J \in \mathcal{J}(\omega, \alpha_+, \alpha_-)$ (see §2.4 for the relevant definitions). Consider an asymptotically cylindrical J-holomorphic curve $u \colon (\dot{\Sigma} = \Sigma \backslash \Gamma, j) \to (\widehat{W}, J)$ asymptotic to nondegenerate[3] Reeb orbits γ_z in $M_\pm$ at its positive/negative punctures $z \in \Gamma^\pm \subset \Sigma$. The index formula for u can be expressed in terms of the Conley–Zehnder indices of its asymptotic orbits, but this requires a choice of normal trivialization along each orbit. We shall therefore fix an arbitrary choice of trivialization of $\gamma^*\xi_\pm$ for every Reeb orbit γ in $M_\pm$ and denote this choice collectively by τ. The Conley–Zehnder index of γ relative to τ will then be denoted by $\mu_{CZ}^\tau(\gamma)$, and we write the **index** of u as

$$\text{ind}(u) := (n - 3)\chi(\dot{\Sigma}) + 2c_1^\tau(u^*T\widehat{W}) + \sum_{z \in \Gamma^+} \mu_{CZ}^\tau(\gamma_z) - \sum_{z \in \Gamma^-} \mu_{CZ}^\tau(\gamma_z), \qquad \text{(A.5)}$$

where $c_1^\tau(u^*T\widehat{W})$ denotes the *relative first Chern number* of the complex vector bundle $(u^*T\widehat{W}, J) \to \dot{\Sigma}$ (cf. §3.4). One can check that the sum on the right-hand side of (A.5) does not depend on the choice τ. As with closed curves, $\text{ind}(u)$

---

[3] Most of this discussion can also be generalized to allow Reeb orbits in Morse–Bott nondegenerate families, though the index formula becomes more complicated (see, e.g., [Bou02, Wen10a]). In general, the linearized Cauchy–Riemann operator is not Fredholm (and thus the moduli space is not well behaved) unless some nondegeneracy condition is imposed on the ends.

A.2 Curves with Punctures 103

is also called the **virtual dimension** of the connected component of $\mathcal{M}_g(\widehat{W}, J)$ containing u; one can show, in fact, that it only depends on the Reeb orbits, the genus, and the relative homology class of u.

A curve $u: (\dot{\Sigma}, j) \to (\widehat{W}, J)$ in $\mathcal{M}_g(\widehat{W}, J)$ is **multiply covered** whenever it can be written as $u = v \circ \varphi$ for some $v: (\dot{\Sigma}', j') \to (\widehat{W}, J)$ in $\mathcal{M}_{g'}(\widehat{W}, J)$ and a holomorphic map

$$\varphi: (\Sigma, j) \to (\Sigma', j') \quad \text{with} \quad \varphi(\dot{\Sigma}) = \dot{\Sigma}',$$

having degree $\deg(\varphi) > 1$. The **automorphism group** $\mathrm{Aut}(\Sigma, j, \Gamma, u)$ can be defined similarly as the group of biholomorphic maps $\varphi: (\Sigma, j) \to (\Sigma, j)$ that fix each point in Γ and satisfy $u = u \circ \varphi$. If u is not multiply covered, it is called **simple**, and then it necessarily has trivial automorphism group. A straightforward combination of standard arguments for the closed case (e.g., [MS12, Prop. 2.5.1]) with Siefring's relative asymptotic formula (Theorem 3.12) proves:

Theorem A.10 *Theorems A.1 and A.2 also hold for asymptotically cylindrical J-holomorphic curves in $\widehat{W}$.*

A proof of the following generalization of Theorem A.3 is sketched in [Wen10a, Theorem 0]:

Theorem A.11 *The subset of $\mathcal{M}_g(\widehat{W}, J)$ consisting of all Fredholm regular curves with trivial automorphism group is open and admits the structure of a smooth finite-dimensional manifold, whose dimension near any given curve $u \in \mathcal{M}_g(\widehat{W}, J)$ is $\mathrm{ind}(u)$.*

We have intentionally omitted the word "oriented" from Theorem A.11, as the question of orientations is somewhat subtler here than in the closed case (see [BM04] or [Wenb, Chapter 11]). Theorem A.4 generalizes as follows:

Theorem A.12 *Assume $\mathcal{U} \subset \widehat{W}$ is an open subset with compact closure, fix $J_0 \in \mathcal{J}(\omega, \alpha_+, \alpha_-)$, and define*

$$\mathcal{J}(\mathcal{U}, J_0) := \left\{ J \in \mathcal{J}(\omega, \alpha_+, \alpha_-) \,\middle|\, J \equiv J_0 \text{ on } \widehat{W} \backslash \mathcal{U} \right\}$$

with its natural C^∞-topology. Then there exists a comeager subset $\mathcal{J}^{\mathrm{reg}}(\mathcal{U}, J_0) \subset \mathcal{J}(\mathcal{U}, J_0)$ such that for all $J \in \mathcal{J}^{\mathrm{reg}}(\mathcal{U}, J_0)$, every simple curve $u \in \mathcal{M}_g(\widehat{W}, J)$ that intersects $\mathcal{U}$ is Fredholm regular.

There is a further variation on the theme of "generic transversality" that only makes sense in the translation invariant setting of a symplectization: a perturbation of a translation-invariant structure $J \in \mathcal{J}(\alpha)$ on $\mathbb{R} \times M$ that is

104 *Properties of Pseudoholomorphic Curves*

generic in the sense of Theorem A.12 cannot generally be assumed translation-invariant, but Dragnev [Dra04] (see also the appendix of [Bou06] or [Wenb, Chapter 8]) proved the following theorem:

Theorem A.13 *Suppose $(M, \xi = \ker \alpha)$ is a closed contact manifold, $\mathcal{U} \subset M$ is an open subset and $J_0 \in \mathcal{J}(\alpha)$, and denote*

$$\mathcal{J}(\mathcal{U}, J_0) := \{J \in \mathcal{J}(\alpha) \mid J \equiv J_0 \text{ on } \mathbb{R} \times (M \backslash \mathcal{U})\}.$$

Then there exists a comeager subset $\mathcal{J}^{\text{reg}}(\mathcal{U}, J_0) \subset \mathcal{J}(\mathcal{U}, J_0)$ such that for all $J \in \mathcal{J}^{\text{reg}}(\mathcal{U}, J_0)$, every simple curve $u \in \mathcal{M}_g(\mathbb{R} \times M, J)$ that intersects $\mathbb{R} \times \mathcal{U}$ is Fredholm regular.

Observe that in the symplectization, the translation invariance of $J \in \mathcal{J}(\alpha)$ turns any curve $u \in \mathcal{M}_g(\mathbb{R} \times M, J)$ that isn't a cover of an orbit cylinder into a 1-parameter family, so Theorem A.13 implies a slightly different analogue of Corollary A.5, as follows:

Corollary A.14 *For generic $J \in \mathcal{J}(\alpha)$ on the symplectization of a closed contact manifold $(M, \xi = \ker \alpha)$, every simple J-holomorphic curve that is not an orbit cylinder satisfies $\text{ind}(u) \geq 1$.*

The punctured generalization of our previous "automatic" transversality result (Theorem A.6) is again valid only in dimension 4, and is most easily stated in terms of the normal Chern number (see §3.4):

Theorem A.15 ([Wen10a, Theorem 1]) *If $\dim \widehat{W} = 4$ and $J \in \mathcal{J}(\omega, \alpha_+, \alpha_-)$, then every immersed J-holomorphic curve $u \in \mathcal{M}_g^A(\widehat{W}, J)$ with $\text{ind}(u) > c_N(u)$ is Fredholm regular.*

Before stating the generalization of Gromov's compactness theorem, we must define the **energy** of a curve $u \in \mathcal{M}_g(\widehat{W}, J)$. The obvious definition (by integrating $u^*\widehat{\omega}$) is not quite the right one, as, for instance, orbit cylinders $u_\gamma(s, t) = (Ts, \gamma(t))$ in the symplectization $(\mathbb{R} \times M, d(e^s \alpha))$ satisfy

$$\int_{\mathbb{R} \times S^1} u_\gamma^* d(e^s \alpha) = \infty.$$

Instead, denote

$$\mathcal{T} := \left\{ \varphi \colon \mathbb{R} \to (-1, 1) \text{ smooth} \mid \varphi'(s) > 0 \text{ for all } s \in \mathbb{R} \text{ and } \varphi(s) \right.$$
$$\left. = s \text{ near } s = 0 \right\},$$

A.2 Curves with Punctures

and observe that for every $\varphi \in \mathcal{T}$, the 2-form on $\widehat{W}$ defined by

$$\omega_\varphi := \begin{cases} \omega & \text{on } W, \\ d\left(e^{\varphi(s)}\alpha_+\right) & \text{on } [0, \infty) \times M_+, \\ d\left(e^{\varphi(s)}\alpha_-\right) & \text{on } (-\infty, 0] \times M_- \end{cases}$$

is symplectic, and any $J \in \mathcal{J}(\omega, \alpha_+, \alpha_-)$ is ω_φ-compatible. We then define

$$E(u) := \sup_{\varphi \in \mathcal{T}} \int_{\dot{\Sigma}} u^* \omega_\varphi \tag{A.6}$$

for any parametrization $u : \dot{\Sigma} \to \widehat{W}$ of a curve in $\mathcal{M}_g(\widehat{W}, J)$.

The natural compactification of $\mathcal{M}_g(\widehat{W}, J)$ is the space $\overline{\mathcal{M}}_g(\widehat{W}, J)$ of **stable J-holomorphic buildings**

$$(v_{N_+}^+, \ldots, v_1^+, v_0, v_1^-, \ldots, v_{N_-}^-),$$

which have $N_+ \geq 0$ **upper levels**, $N_- \geq 0$ **lower levels** and exactly one **main level**. Each of the levels is a (possibly disconnected) asymptotically cylindrical nodal curve that is stable in the sense defined in §A.1, where

- v_i^+ for $i = 1, \ldots, N_+$ live in $\mathbb{R} \times M_+$ and are J_+-holomorphic, with

$$J_+ := J|_{[0,\infty) \times M_+} \in \mathcal{J}(\alpha_+).$$

- v_0 lives in $\widehat{W}$ and is J-holomorphic.
- v_i^- for $i = 1, \ldots, N_-$ live in $\mathbb{R} \times M_-$ and are J_--holomorphic, with

$$J_- := J|_{(-\infty, 0] \times M_-} \in \mathcal{J}(\alpha_-).$$

The levels also connect to each other, meaning that the data of a building include a bijection between the positive ends of each level and the negative ends of the level above it such that matching ends are asymptotic to the same Reeb orbit – orbits that appear in this way are not considered asymptotic orbits of the building itself, but are sometimes called **breaking orbits** (see Figure A.3). The **arithmetic genus** $g \geq 0$ can be characterized by the following condition: if $\widehat{S}$ denotes the space obtained from the domains of all the levels by filling in all punctures and then identifying any two nodal points that belong to the same node and any two punctures between levels that are matched by the aforementioned bijection, then $\widehat{S}$ is homeomorphic to a (possibly singular) fiber of some Lefschetz fibration with closed regular fibers of genus g. Equivalence of holomorphic buildings is defined via the obvious notion of biholomorphic equivalence (preserving nodes and matching punctures), with the additional

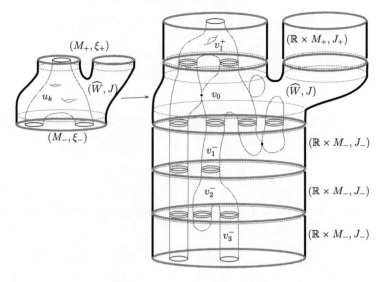

Figure A.3 Degeneration of a sequence u_k of punctured holomorphic curves with genus 2, one positive end and two negative ends in a symplectic cobordism. The limiting holomorphic building $(v_1^+, v_0, v_1^-, v_2^-, v_3^-)$ in this example has one upper level, a main level and three lower levels, each of which is a (possibly disconnected) punctured nodal holomorphic curve. The building has arithmetic genus 2 and the same numbers of positive and negative ends as u_k.

feature that upper and lower levels may be translated freely; i.e., two levels that are identical after an $\mathbb{R}$-translation of an upper or lower level are considered equivalent. Finally, the stability condition is enhanced with the stipulation that none of the levels $v_i^\pm$ may consist exclusively of orbit cylinders without any nodes; this is necessary in order to make sure that the natural topology of $\overline{\mathcal{M}}_g(\widehat{W}, J)$ is Hausdorff.

The natural inclusion

$$\mathcal{M}_g(\widehat{W}, J) \hookrightarrow \overline{\mathcal{M}}_g(\widehat{W}, J)$$

regards any smooth curve $u \in \mathcal{M}_g(\widehat{W}, J)$ as a building that has no upper or lower levels and no nodes.

Theorem A.16 *For every $g \geq 0$ and every $J \in \mathcal{J}(\omega, \alpha_+, \alpha_-)$, $\overline{\mathcal{M}}_g(\widehat{W}, J)$ admits a natural topology as a metrizable space, and its connected components are compact. Moreover, any sequence $u_k \in \mathcal{M}_g(\widehat{W}, J)$ of curves satisfying a uniform energy bound $E(u_k) \leq C$ in the sense of (A.6) has a subsequence convergent to an element of $\overline{\mathcal{M}}_g(\widehat{W}, J)$.*

A.2 Curves with Punctures 107

A small modification is appropriate in the case where $(\widehat{W}, J)$ is the completion of a *trivial* symplectic cobordism, i.e., an $\mathbb{R}$-invariant symplectization $\mathbb{R} \times M$. In this case, the levels are still ordered, but there is no distinguished main level, nor a distinction between "upper" and "lower" levels, and the notion of equivalence allows $\mathbb{R}$-translations in all levels – the latter means in particular that $\overline{\mathcal{M}}_g(\mathbb{R} \times M, J)$ is not a compactification of $\mathcal{M}_g(\mathbb{R} \times M, J)$, but rather of $\mathcal{M}_g(\mathbb{R} \times M, J)/\mathbb{R}$. For full details on these matters, including a precise definition of the notion of convergence to a holomorphic building, we refer to [BEH$^+$03].

Appendix B

Local Positivity of Intersections

In this appendix we explain the local results in the background of the standard theorems of §2.1 on positivity of intersections and the adjunction formula. Readers wishing to understand the geometric picture without worrying about the analytical details may read the statement of Theorem B.23 in §B.2 and then skip ahead to §B.3, which proves positivity of intersections using the local representation formula of Theorem B.23 as a black box. The main tool in proving the latter is the similarity principle, which is explained (along with the necessary background on elliptic regularity) in §B.1.

Since all important results in this appendix are local, we will mostly discuss functions defined on the domains

$$\mathbb{D} := \left\{ z \in \mathbb{C} \,\middle|\, |z| \le 1 \right\} \quad \text{and} \quad \mathbb{D}_\rho := \left\{ z \in \mathbb{C} \,\middle|\, |z| \le \rho \right\}$$

for $\rho > 0$.

B.1 Regularity and the Similarity Principle

The similarity principle can be thought of as a linearized version of positivity of intersections: it gives a local description of solutions to linear Cauchy–Riemann type equations near their zeroes, proving in particular that they qualitatively resemble complex-analytic functions. The proof given in this section is more or less self-contained – it requires some understanding of the theory of distributions and Sobolev spaces, but avoids using the harder aspects of elliptic regularity theory such as the Calderón–Zygmund inequality. It is based in large part on arguments that were explained to the author by Jean-Claude Sikorav.

B.1.1 Linear Cauchy–Riemann Type Operators

Linear Cauchy–Riemann equations on vector bundles arise naturally from infinitesimal perturbations of J-holomorphic curves.

Definition B.1 Suppose (Σ, j) is a Riemann surface, $E \to \Sigma$ is a smooth complex vector bundle, and $F \to \Sigma$ denotes the complex vector bundle

$$F := \overline{\mathrm{Hom}}_{\mathbb{C}}(T\Sigma, E)$$

whose sections are the complex-antilinear bundle maps $T\Sigma \to E$. A (smooth) **linear Cauchy–Riemann type operator** is a first-order real-linear partial differential operator $\mathbf{D} \colon \Gamma(E) \to \Gamma(F)$ that satisfies the Leibniz rule

$$\mathbf{D}(f\eta) = (\bar{\partial}f)\eta + f\mathbf{D}\eta \quad \text{for all} \quad \eta \in \Gamma(E),\, f \in C^\infty(\Sigma, \mathbb{R}),$$

where $\bar{\partial}f \in \Omega^{0,1}(T\Sigma)$ denotes the complex-valued 1-form $df + i\, df \circ j$.

Remark B.2 If $\mathbf{D} \colon \Gamma(E) \to \Gamma(F)$ in the above definition is also complex linear, then it satisfies a complex version of the Leibniz rule, namely

$$\mathbf{D}(f\eta) = (\bar{\partial}f)\eta + f\mathbf{D}\eta \quad \text{for all} \quad \eta \in \Gamma(E),\, f \in C^\infty(\Sigma, \mathbb{C}).$$

It is important, however, to allow the possibility that $\mathbf{D} \colon \Gamma(E) \to \Gamma(F)$ is only real and not complex linear, even though E and F both carry complex structures. Unless one restricts attention to complex manifolds with *integrable* complex structures, most of the linearized Cauchy–Riemann operators that arise in the context of J-holomorphic curve theory are not complex linear.

Remark B.3 It is easy to check that if $\mathbf{D} \colon \Gamma(E) \to \Gamma(F)$ is a Cauchy–Riemann type operator and $A \colon E \to F$ is a smooth real-linear bundle map, then $\mathbf{D} + A$ is also a Cauchy–Riemann type operator. Moreover, for any two Cauchy–Riemann type operators $\mathbf{D}$ and $\mathbf{D}'$ on E, the map $\mathbf{D}' - \mathbf{D} \colon \Gamma(E) \to \Gamma(F)$ is C^∞-linear and thus arises from a smooth bundle map $A \colon E \to F$, meaning $\mathbf{D}' = \mathbf{D} + A$. This proves that the space of all Cauchy–Riemann type operators on E is an affine space over $\Gamma(\mathrm{Hom}_{\mathbb{R}}(E, F))$.

Given an open subset $\mathcal{U} \subset \Sigma$ with a holomorphic coordinate $z = s + it \colon \mathcal{U} \to \mathbb{C}$ identifying $(\mathcal{U}, j)$ with $(\mathbb{D}, i)$ and a complex trivialization of $E|_{\mathcal{U}}$, there is a naturally induced trivialization of $F|_{\mathcal{U}}$ such that if $\eta \in \Gamma(E|_{\mathcal{U}})$ is represented by the function $f \colon \mathbb{D} \to \mathbb{C}^n$, then the same function also represents the section $\xi \in \Gamma(F|_{\mathcal{U}})$ given by $\xi(X) = d\bar{z}(X)\,\eta(p)$ for $p \in \mathcal{U}$ and $X \in T_p\Sigma$. These choices identify the spaces of sections of E and F over $\mathcal{U}$ with $C^\infty(\mathbb{D}, \mathbb{C}^n)$ such that the map

$$\partial_s + i\partial_t \colon C^\infty(\mathbb{D}, \mathbb{C}^n) \to C^\infty(\mathbb{D}, \mathbb{C}^n)$$

110 *Local Positivity of Intersections*

represents a linear Cauchy–Riemann type operator on $E|_{\mathcal{U}}$. It follows via Remark B.3 that *every* Cauchy–Riemann type operator $\mathbf{D}\colon \Gamma(E|_{\mathcal{U}}) \to \Gamma(F|_{\mathcal{U}})$ is in this way identified with a map of the form

$$\partial_s + i\partial_t + A\colon C^\infty(\mathbb{D}, \mathbb{C}^n) \to C^\infty(\mathbb{D}, \mathbb{C}^n) \tag{B.1}$$

for some smooth function $A\colon \mathbb{D} \to \mathrm{End}_{\mathbb{R}}(\mathbb{C}^n)$. With this local picture understood, it will sometimes also be useful to consider Cauchy–Riemann type operators on complex vector bundles that are not equipped with a smooth structure, e.g., pullbacks of smooth bundles along nonsmooth (but differentiable) maps. In general, one says that a vector bundle is **of class** C^k if it is equipped with an atlas of local trivializations whose transition maps are all of class C^k. One can then speak of sections of class C^m for any $m \le k$ but not for $m > k$; the former notion makes sense due to the fact that for $m \le k$, the product of a C^k-smooth function with a C^m-smooth function is also of class C^m. In the following, we shall allow nonsmooth vector bundles $E \to \Sigma$ but continue to assume that the base is a *smooth* Riemann surface; i.e., the almost complex structure j on Σ is smooth, so that holomorphic local coordinate charts on Σ are automatically also smooth. For this reason, $F := \overline{\mathrm{Hom}}_{\mathbb{C}}(T\Sigma, E)$ always inherits from E and Σ an atlas of local trivializations with the same regularity as E. If E is of class C^k, then the notion of a differential operator from E to F of order $r \in \mathbb{N}$ makes sense as long as $r \le k$.

Definition B.4 Suppose (Σ, j) is a Riemann surface, $E \to \Sigma$ is a complex vector bundle of class C^k for some $k \in \mathbb{N} \cup \{\infty\}$, $F = \overline{\mathrm{Hom}}_{\mathbb{C}}(T\Sigma, E)$, and $m \le k - 1$ is a nonnegative integer. A **linear Cauchy–Riemann type operator of class** C^m is a first-order real-linear partial differential operator $\mathbf{D}$ from E to F such that under arbitrary choices of local holomorphic coordinates and trivializations as described in the previous paragraph, $\mathbf{D}$ locally takes the form $\partial_s + i\partial_t + A$ for some $A \in C^m(\mathbb{D}, \mathrm{End}_{\mathbb{R}}(\mathbb{C}^n))$.

Remark B.5 The condition $m \le k - 1$ is required in the above definition since the transformation of the zeroth-order term in a Cauchy–Riemann type operator under a transition map depends in general on the first derivative of the transition map; i.e., if the latter is only of class C^k, then the condition $A \in C^{k-1}$ is coordinate invariant, but $A \in C^k$ would not be. The same remark applies to connections on a bundle of class C^k.

Remark B.6 For functions of Sobolev class $W^{k,p}$, there is also a well-defined continuous product pairing $C^k \times W^{k,p} \to W^{k,p}$ due to the fact that products of continuous functions with L^p-functions are also in L^p. As a consequence, one can also speak of Cauchy–Riemann type operators of class $W^{k,p}$ whenever the bundle is of class C^{k+1}.

B.1 Regularity and the Similarity Principle 111

Example B.7 If $E \to \Sigma$ is endowed with a holomorphic vector bundle structure, then it carries a canonical (complex-)linear Cauchy–Riemann type operator $\mathbf{D}\colon \Gamma(E) \to \Gamma(F)$ such that the local holomorphic functions $\eta \in \Gamma(E|_\mathcal{U})$ on open sets $\mathcal{U} \subset \Sigma$ are precisely those that satisfy $\mathbf{D}\eta = 0$. This operator takes the form $\partial_s + i\partial_t$ with respect to any choice of local holomorphic coordinates and holomorphic trivialization, and the holomorphicity of the transition maps guarantees that this definition does not depend on any choices.

Example B.8 For any connection ∇ on $E \to \Sigma$, $\mathbf{D}\eta := \nabla\eta + i\,\nabla\eta \circ j$ defines a linear Cauchy–Riemann type operator.

Example B.9 If $u\colon (\Sigma, j) \to (M, J)$ is a J-holomorphic curve, then linearizing the nonlinear operator $\bar{\partial}_J(u) := du + J \circ du \circ j$ along a smooth family of maps $\{u_\sigma\colon \Sigma \to M\}_{\sigma \in (-\epsilon, \epsilon)}$ with $u_0 = u$ and $\eta := \partial_\sigma u_\sigma|_{\sigma=0} \in \Gamma(u^*TM)$ gives rise to a linear Cauchy–Riemann type operator of the form

$$\Gamma(u^*TM) \overset{\mathbf{D}_u}{\to} \Gamma(\overline{\mathrm{Hom}_{\mathbb{C}}(T\Sigma, u^*TM)}),$$

$$\eta \mapsto \nabla\eta + J(u) \circ \nabla\eta \circ j + (\nabla_\eta J) \circ Tu \circ j,$$

where ∇ is an arbitrary choice of symmetric connection on M.

B.1.2 Elliptic Regularity

In the following discussion, we will consider $\mathbb{C}^n$-valued functions of one complex variable $z = s + it$, for which we denote the standard local models of Cauchy–Riemann and anti-Cauchy–Riemann type operators by

$$\bar{\partial} := \partial_s + i\partial_t, \qquad \partial := \partial_s - i\partial_t.$$

Notation (Sobolev spaces). In this appendix, the Sobolev space of functions $f\colon \mathbb{D} \to \mathbb{C}^n$ admitting weak derivatives of class L^p up to order $k \geq 0$ is denoted by $W^{k,p}(\mathbb{D})$, and for the case $p = 2$ we abbreviate the Hilbert spaces $H^k(\mathbb{D}) := W^{k,2}(\mathbb{D})$. The larger vector spaces $W^{k,p}_{\mathrm{loc}}(\mathbb{D})$ and $H^k_{\mathrm{loc}}(\mathbb{D})$ consist of all functions on $\mathbb{D}$ whose restrictions to compact subsets of the interior of $\mathbb{D}$ are of class $W^{k,p}$ or H^k, respectively. An important special case of this is $L^1_{\mathrm{loc}}(\mathbb{D}) = W^{0,1}_{\mathrm{loc}}(\mathbb{D})$, the space of all locally integrable functions on $\mathbb{D}$. We write the space of smooth compactly supported functions on the interior of $\mathbb{D}$ as $C^\infty_0(\mathbb{D})$.

Since $\bar{\partial}$ and ∂ are first-order differential operators with constant coefficients, they define bounded linear maps

$$\bar{\partial}, \partial\colon W^{k,p}(\mathbb{D}) \to W^{k-1,p}(\mathbb{D})$$

112   *Local Positivity of Intersections*

for each $k \in \mathbb{N}$ and $p \in [1, \infty]$. We will need to use the "easy" ($p = 2$) case of the following nontrivial fact from elliptic regularity theory.

Proposition B.10   *For each $p \in (1, \infty)$, the operator $\bar{\partial} \colon W^{1,p}(\mathbb{D}) \to L^p(\mathbb{D})$ is surjective and admits a bounded right inverse $T \colon L^p(\mathbb{D}) \to W^{1,p}(\mathbb{D})$.*

Sketch of the proof for $p = 2$ The locally integrable function $K \colon \mathbb{C} \to \mathbb{C}$ defined almost everywhere by

$$K(z) := \frac{1}{2\pi z}$$

is a fundamental solution for the equation $\bar{\partial} u = f$, meaning it satisfies $\bar{\partial} K = \delta$ in the sense of distributions, so, in particular, one can use convolution to associate to any $f \in C_0^\infty(\mathbb{C})$ a function $u := K * f \in C^\infty(\mathbb{C})$ satisfying $\bar{\partial} u = f$. Since $C_0^\infty(\mathbb{D}) \subset C_0^\infty(\mathbb{C})$ is dense in $L^p(\mathbb{D})$, the desired right inverse $T \colon L^p(\mathbb{D}) \to W^{1,p}(\mathbb{D})$ can be defined as the unique bounded linear extension of $C_0^\infty(\mathbb{D}) \to C^\infty(\mathbb{D}) \colon f \mapsto (K * f)|_{\mathbb{D}}$ after establishing an estimate of the form

$$\|K * f\|_{W^{1,p}(\mathbb{D})} \le c\|f\|_{L^p} \quad \text{for all} \quad f \in C_0^\infty(\mathbb{D}).$$

This is equivalent to three estimates,

$$\|K * f\|_{L^p(\mathbb{D})} \le c\|f\|_{L^p}, \quad \|\partial(K * f)\|_{L^p(\mathbb{D})} \le c\|f\|_{L^p},$$
$$\|\bar{\partial}(K * f)\|_{L^p(\mathbb{D})} \le c\|f\|_{L^p}, \tag{B.2}$$

each again for $f \in C_0^\infty(\mathbb{D})$. The third of these is immediate since $\bar{\partial}(K * f) = f$. The first estimate is a minor variation on the standard *Young's inequality* for convolutions (see, e.g., [LL01, §4.2]), and admits a similar proof based on the Hölder inequality and Fubini's theorem – the crucial assumptions here are only that K is locally integrable and $\mathbb{D} \subset \mathbb{C}$ is bounded. The hard part in general is the second estimate, though in the case $p = 2$, a straightforward argument is possible using the Fourier transform.

The idea is to interpret both sides of the equation $\bar{\partial} K = \delta$ as tempered distributions on $\mathbb{C}$, which then have well-defined Fourier transforms in the sense of distributions. Expressing these Fourier transforms as functions of a variable $\zeta \in \mathbb{C}$, the Fourier transform $\widehat{K}(\zeta)$ of $K(z)$ gets multiplied by $2\pi i \zeta$ to produce the Fourier transform of $\bar{\partial} K(z)$, so $\bar{\partial} K = \delta$ implies

$$2\pi i \zeta \widehat{K}(\zeta) = \widehat{\delta}(\zeta) = 1. \tag{B.3}$$

If $f \in C_0^\infty(\mathbb{C})$ and we define another function on $\mathbb{C}$ by $u := K * f$, then u also defines a tempered distribution, whose Fourier transform $\widehat{u}$ then satisfies

$$\widehat{u} = \widehat{K * f} = \widehat{K}\widehat{f},$$

B.1 Regularity and the Similarity Principle 113

so by (B.3) we have $2\pi i \zeta \widehat{u}(\zeta) = \widehat{f}(\zeta)$. Denoting the Lebesgue measure on $\mathbb{C}$ for functions of $\zeta \in \mathbb{C}$ by $d\mu(\zeta)$, Plancherel's theorem now implies

$$
\begin{aligned}
\|\partial(K * f)\|_{L^2(\mathbb{D})}^2 \le \|\partial u\|_{L^2(\mathbb{C})}^2 &= \int_{\mathbb{C}} \left|\widehat{\partial u}(\zeta)\right|^2 d\mu(\zeta) = \int_{\mathbb{C}} \left|2\pi i \overline{\zeta} \widehat{u}(\zeta)\right|^2 d\mu(\zeta) \\
&= \int_{\mathbb{C}} \left|\frac{\overline{\zeta}}{\zeta} 2\pi i \zeta \widehat{u}(\zeta)\right|^2 d\mu(\zeta) = \int_{\mathbb{C}} \left|\widehat{f}(\zeta)\right|^2 d\mu(\zeta) = \|\widehat{f}\|_{L^2(\mathbb{C})}^2 \\
&= \|f\|_{L^2(\mathbb{C})}^2,
\end{aligned}
$$

and the last expression is the same as $\|f\|_{L^2(\mathbb{D})}^2$ if we assume $f \in C_0^\infty(\mathbb{D})$; hence the remaining estimate is proven.

The case $p \ne 2$ requires totally different arguments, which begin by writing $\partial(K * f) = \partial K * f$ as a principal value integral

$$
\partial(K * f)(z) = -\frac{1}{\pi} \lim_{\epsilon \to 0^+} \int_{|\zeta - z| \ge \epsilon} \frac{f(\zeta)}{(z - \zeta)^2} d\mu(\zeta),
$$

in which the right-hand side can be interpreted as the convolution of a distribution ∂K with a smooth function f. The limit in this expression is necessary because in contrast to $1/z$, the function $1/z^2$ that arises by differentiating $K(z)$ is not locally integrable on $\mathbb{C}$, and, for this reason, simple convolution inequalities do not apply. Estimates in L^p for transformations given by singular integrals of this type are the subject of a much harder analytical result, the Calderón–Zygmund inequality. Details on this and the rest of the argument sketched above may be found in [Wena, Chapter 2]; we shall not present them here since the $p \ne 2$ case, while important for the general theory of pseudoholomorphic curves, is not needed in our discussion of intersection theory. $\qquad\square$

Remark B.11 A closely related result is the existence of an estimate

$$
\|u\|_{W^{1,p}} \le c \|\bar{\partial} u\|_{L^p} \quad \text{for all} \quad u \in C_0^\infty(\mathbb{D}). \tag{B.4}
$$

This can be derived from Proposition B.10 using a bit of extra knowledge about the fundamental solution $K(z) = 1/2\pi z$, but in the case $p = 2$ it also admits the following simple proof borrowed from Sikorav [Sik94]. The bound on $\|u\|_{W^{1,p}}$ is again equivalent to three bounds, namely on $\|u\|_{L^p}$, $\|\partial u\|_{L^p}$ and $\|\bar{\partial} u\|_{L^p}$, where the third is immediate. Since u is assumed to be smooth with compact support, the first bound follows from a standard Sobolev estimate, the Poincaré inequality (see, e.g., [AF03, §6.30]). To achieve the second bound, it is convenient to write $z = s + it \in \mathbb{C}$ and consider the complex partial derivative operators

$$
\partial_z := \frac{\partial}{\partial z} = \frac{1}{2}\partial, \qquad \partial_{\bar{z}} := \frac{\partial}{\partial \bar{z}} = \frac{1}{2}\bar{\partial}
$$

114 *Local Positivity of Intersections*

along with the corresponding complex-valued 1-forms

$$dz = ds + i\,dt, \qquad d\bar{z} = ds - i\,dt.$$

For any smooth compactly supported function $u\colon \mathbb{C} \to \mathbb{C}$, we can now write

$$du = \partial_z u\,dz + \partial_{\bar{z}} u\,d\bar{z},$$

and the complex-valued 1-form $u\,d\bar{u}$ has compact support in $\mathbb{C}$, so applying Stokes's theorem to $d(u\,d\bar{u}) = du \wedge d\bar{u}$ on a sufficiently large disk $\mathbb{D}_R \subset \mathbb{C}$ gives

$$0 = \int_{\partial \mathbb{D}_R} u\,d\bar{u} = \int_{\mathbb{D}_R} du \wedge d\bar{u} = \int_{\mathbb{D}_R} (\partial_z u\,dz + \partial_{\bar{z}} u\,d\bar{z}) \wedge (\partial_z \bar{u}\,dz + \partial_{\bar{z}} \bar{u}\,d\bar{z})$$

$$= \frac{1}{4} \int_{\mathbb{D}_R} \left(|\partial u|^2 - |\bar{\partial} u|^2 \right) dz \wedge d\bar{z},$$

proving $\|\partial u\|_{L^2} = \|\bar{\partial} u\|_{L^2}$.

Note that by applying (B.4) to derivatives $\partial^\alpha u$ with arbitrary multi-indices α and using the fact that ∂^α commutes with $\bar{\partial}$, one obtains the easy generalization

$$\|u\|_{W^{k,p}} \le c\|\bar{\partial} u\|_{W^{k-1,p}} \qquad \text{for all} \quad u \in C_0^\infty(\mathbb{D})$$

for every $k \in \mathbb{N}$. By density, this extends to

$$\|u\|_{W^{k,p}} \le c\|\bar{\partial} u\|_{W^{k-1,p}} \qquad \text{for all} \quad u \in W_0^{k,p}(\mathbb{D}), \tag{B.5}$$

where $W_0^{k,p}(\mathbb{D}) \subset W^{k,p}(\mathbb{D})$ denotes the closed subspace defined as the $W^{k,p}$-closure of $C_0^\infty(\mathbb{D})$.

Note that if $p > 2$ and $f \in L^p(\mathbb{D})$, then the statement of Proposition B.10 gives a solution $u := Tf \in W^{1,p}(\mathbb{D})$ to the equation $\bar{\partial} u = f$, and u is then *continuous* by the Sobolev embedding theorem. We will need this fact for certain applications, but since we did not prove the $p > 2$ case of Proposition B.10, the continuity of solutions to $\bar{\partial} u = f \in L^p$ for $p > 2$ needs to be proved separately. This turns out to be not so hard.

Proposition B.12 *Let $T\colon L^2(\mathbb{D}) \to H^1(\mathbb{D})$ denote the bounded right inverse of $\bar{\partial}\colon H^1(\mathbb{D}) \to L^2(\mathbb{D})$ provided by Proposition B.10, defined as an extension of the convolution operator $f \mapsto K * f$. Then for $p \in (2, \infty)$, T sends any $f \in L^p(\mathbb{D}) \subset L^2(\mathbb{D})$ into $C^0(\mathbb{D})$, and it restricts to a bounded linear operator*

$$T\colon L^p(\mathbb{D}) \to C^0(\mathbb{D}).$$

Proof Observe that the fundamental solution $K(z) = 1/2\pi z$ belongs to $L^q_{\text{loc}}(\mathbb{C})$ whenever $1 \le q < 2$, and, in fact, the L^q-norm of K on the unit disk $\mathbb{D}(z) \subset \mathbb{C}$ about a point $z \in \mathbb{C}$ satisfies

$$\|K\|_{L^q(\mathbb{D}(z))} \le C$$

B.1 Regularity and the Similarity Principle

for some constant $C > 0$ that depends on q but not on z. In particular, if $p > 2$, this is true for $q \in (1, 2)$ such that $1/q + 1/p = 1$. Now if $f \in C_0^\infty(\mathbb{D})$, Hölder's inequality implies that for every $z \in \mathbb{C}$,

$$|K * f(z)| = \left| \int_{\mathbb{C}} K(z - \zeta) f(\zeta)\, d\mu(\zeta) \right| \le \int_{\mathbb{D}} |K(z - \zeta)| \cdot |f(\zeta)|\, d\mu(\zeta)$$

$$\le \|K(z - \cdot)\|_{L^q(\mathbb{D})} \cdot \|f\|_{L^p(\mathbb{D})} \le C\|f\|_{L^p(\mathbb{D})};$$

hence $\|Tf\|_{L^\infty} \le C\|f\|_{L^p}$. By standard results on convolutions of smooth functions with distributions (see, e.g., [LL01, §6.13]), $K * f$ is a smooth function for each $f \in C_0^\infty(\mathbb{D})$, and thus the map $f \mapsto K * f$ extends to a bounded linear map from $L^p(\mathbb{D})$ to the L^∞-closure of the space of bounded smooth functions, which is $C^0(\mathbb{D})$. Since $L^p(\mathbb{D})$ embeds continuously into $L^2(\mathbb{D})$, this extension is necessarily the same as $T : L^2(\mathbb{D}) \to H^1(\mathbb{D})$ for all $f \in L^p(\mathbb{D})$. $\qquad\square$

Here is the first of several applications of these estimates:

Proposition B.13 *If $g \in H_{\mathrm{loc}}^k(\mathbb{D})$ for some integer $k \ge 0$, then every weak solution $f \in L_{\mathrm{loc}}^1(\mathbb{D})$ to the equation $\bar\partial f = g$ is also in $H_{\mathrm{loc}}^{k+1}(\mathbb{D})$. In particular, f is smooth whenever g is smooth.*

Sketch of the proof One starts by showing that if $k \ge 1$ and f is already known to be in $H^1(\mathbb{D}_r)$ for some $r > 0$, then f will also be in $H^{k+1}(\mathbb{D}_{r'})$ for every $r' \in (0, r)$, and it satisfies an estimate of the form

$$\|f\|_{H^{k+1}(\mathbb{D}_{r'})} \le c\|f\|_{H^1(\mathbb{D}_r)} + c\|g\|_{H^k(\mathbb{D}_r)}. \tag{B.6}$$

To show, for instance, that f is in $H^2(\mathbb{D}_{r'})$, it suffices to show that both of the partial derivatives $\partial_s f$ and $\partial_t f$ are in $H^1(\mathbb{D}_{r'})$, and for this purpose one can approximate them by difference quotients, e.g.,

$$D_s^h f(s, t) := \frac{f(s + h, t) - f(s, t)}{h}$$

for $h \in \mathbb{R}\backslash\{0\}$ close enough to 0 so that this definition makes sense on a neighborhood of $\mathbb{D}_{r'}$. These difference quotients are automatically of class H^1 since f is, and the main task is to show that they satisfy a uniform H^1-bound on $\mathbb{D}_{r'}$ as $h \to 0$, as the Banach–Alaoglu theorem then implies that they converge weakly to a function in H^1 as $h \to 0$, implying that $\partial_s f$ (or $\partial_t f$, respectively) is indeed of class H^1. The required uniform bound comes from the basic elliptic estimate (B.5). To apply it, one chooses a smooth function $\beta : \mathbb{D}_r \to [0, 1]$ that equals 1 on $\mathbb{D}_{r'}$ and has compact support in the interior of $\mathbb{D}_r$, so that $\beta D_s^h f$ is now of class H_0^1 on $\mathbb{D}_r$; thus $\|D_s^h f\|_{H^1(\mathbb{D}_{r'})} \le \|\beta D_s^h f\|_{H^1(\mathbb{D}_r)}$ is bounded in terms of $\|\bar\partial(\beta D_s^h f)\|_{L^2(\mathbb{D}_r)}$. This, in turn, can be bounded in terms of $\|D_s^h f\|_{L^2}$

116   *Local Positivity of Intersections*

and $\|D_s^h g\|_{L^2}$, which are both uniformly bounded as $h \to 0$ because f and g are both of class H^1. If $k > 1$, then one can now repeat this argument with the knowledge that f is of class H^2 on $\mathbb{D}_{r'}$, and continue repeating it on smaller disks at each step until f is shown to be of class H^{k+1}, with the estimate (B.6) as a quantitative expression of this fact. This argument shows, in particular, that if f is of class H^1_{loc} and $\bar{\partial} f$ is of class H^k_{loc}, then f is also of class H^{k+1}_{loc}.

Before weakening the hypothesis on f further, it is useful to notice that the previous paragraph makes possible a generalization of Proposition B.10: for every $k \in \mathbb{N}$, the operator $\bar{\partial} \colon H^k(\mathbb{D}) \to H^{k-1}(\mathbb{D})$ admits a bounded right inverse

$$T_k \colon H^{k-1}(\mathbb{D}) \to H^k(\mathbb{D}).$$

The proof of this is by induction on k, with Proposition B.10 as the case $k = 0$. If one fixes some $R > 1$ and assumes that a right inverse $T_{k-1} \colon H^{k-2}(\mathbb{D}_R) \to H^{k-1}(\mathbb{D}_R)$ of $\bar{\partial} \colon H^{k-1}(\mathbb{D}_R) \to H^{k-2}(\mathbb{D}_R)$ exists, then a right inverse $T_k \colon H^{k-1}(\mathbb{D}) \to H^k(\mathbb{D})$ for $\bar{\partial} \colon H^k(\mathbb{D}) \to H^{k-1}(\mathbb{D})$ can be defined in the form

$$T_k f := T_{k-1} \widetilde{f} \big|_{\mathbb{D}}$$

for $f \in H^{k-1}(\mathbb{D}) \subset H^{k-2}(\mathbb{D})$, where

$$H^{k-1}(\mathbb{D}) \to H^{k-1}(\mathbb{D}_R) \colon f \mapsto \widetilde{f}$$

is any choice of bounded linear extension operator, i.e., satisfying $\widetilde{f}|_{\mathbb{D}} = f$. The reason this defines a bounded operator $H^{k-1}(\mathbb{D}) \to H^k(\mathbb{D})$ is that, if $u = T_k f$, then u is the restriction to a smaller disk $\mathbb{D} \subset \mathbb{D}_R$ of a function $T_{k-1}\widetilde{f} \in H^{k-1}(\mathbb{D}_R)$ that satisfies $\bar{\partial} T_{k-1}\widetilde{f} = \widetilde{f} \in H^{k-1}(\mathbb{D}_R)$; thus the previous paragraph implies that u is also in $H^k(\mathbb{D})$, and (B.6) produces the required estimate on $\|u\|_{H^k(\mathbb{D})}$.

Finally, if $f \in L^1(\mathbb{D}_r)$ and $\bar{\partial} f = g \in H^k(\mathbb{D}_r)$, one can now take the bounded right inverse $T_{k+1} \colon H^k(\mathbb{D}_r) \to H^{k+1}(\mathbb{D}_r)$ and consider the function $h := f - T_{k+1}g$, which is in $L^1(\mathbb{D}_r)$ and is a weak solution to the equation $\bar{\partial} h = 0$. The real and imaginary parts of h are then weak solutions to the Laplace equation, and, by convolution with an approximate identity, one can approximate them in $L^1(\mathbb{D}_r)$ by smooth solutions to the Laplace equation. The latter are characterized by the mean value property (see [Eva98, §2.2.3]), which behaves well under L^1-convergence, implying that the real and imaginary parts of h also satisfy the mean value property and are therefore smooth. In particular, h then belongs to $H^{k+1}_{\text{loc}}(\mathbb{D}_r)$, and therefore so does $f = h + T_{k+1}g$. $\square$

B.1 Regularity and the Similarity Principle 117

Corollary B.14 *Suppose E is a complex vector bundle of class C^{k+1} over a Riemann surface Σ, and $\mathbf{D}: \Gamma(E) \to \Gamma(\overline{\mathrm{Hom}}_{\mathbb{C}}(T\Sigma, E))$ is a linear Cauchy–Riemann type operator of class C^k. Then every weak solution of class L^2_{loc} to the equation $\mathbf{D}\eta = 0$ is of class H^{k+1}_{loc}. In particlar, if the bundle E and operator $\mathbf{D}$ are smooth, then all weak solutions of class L^2_{loc} are smooth.*

Proof Locally, a weak solution to $\mathbf{D}\eta = 0$ of class L^2_{loc} can be represented by a $\mathbb{C}^n$-valued function $f \in L^2(\mathbb{D})$ satisfying $(\bar{\partial} + A)f = 0$ for some function $A: \mathbb{D} \to \mathrm{End}_{\mathbb{R}}(\mathbb{C}^n)$ of class C^k. In particular, A is continuous, so $-Af$ is of class L^2, and the equation $\bar{\partial}f = -Af$ thus implies via Proposition B.13 that f is of class H^1_{loc}. If A is also of class C^1, it follows that $-Af$ is in H^1_{loc}, and another application of Proposition B.13 implies $f \in H^2_{\mathrm{loc}}$. Repeat until A runs out of derivatives. □

At one point in §B.3 we will need a nonlinear analogue of the foregoing result, which applies to J-holomorphic curves in an almost complex manifold (M, J). This justifies the fact that we only consider *smooth* J-holomorphic curves in this book, even though the nonlinear Cauchy–Riemann equation would make sense for maps that are only differentiable. The hypotheses can be weakened in various ways, e.g., by allowing nonsmooth almost complex structures, but we will have no need to consider this. The statement is fundamentally local, and thus we are free to assume $(M, J) = (\mathbb{C}^n, J)$.

Proposition B.15 *Suppose J is a smooth almost complex structure on $\mathbb{C}^n$ with $J(0) = i$ and $u: \mathbb{D} \to \mathbb{C}^n$ is a continuous function of class $W^{1,\infty}$ that is a weak solution to the equation $\partial_s u + J(u)\partial_t u = 0$ with $u(0) = 0$. Then u is smooth.*

Sketch of the proof By the Sobolev embedding theorem, it suffices to prove that u is of class H^k_{loc} for every $k \in \mathbb{N}$. We prove this by induction on k, and observe first that, at each step of the induction, it will be enough to prove that u is of class H^k on $\mathbb{D}_\rho$ for some $\rho > 0$; indeed, changing coordinates then produces the same result on sufficiently small neighborhoods of any point in the domain, so that finitely many such small neighborhoods can be pieced together to show that u is in H^k on any compact subset of the interior.

Another useful observation is that, for any constants $R > 0$ and $\rho \in (0, 1]$, u is of class H^k on $\mathbb{D}_\rho$ if and only if the rescaled map

$$\hat{u}: \mathbb{D} \to \mathbb{C}^n : z \mapsto Ru(\rho z)$$

is in $H^k(\mathbb{D})$. To make use of this, we rewrite the equation $\partial_s u + J(u)\partial_t u = 0$ in the form

$$\bar{\partial}u - Q(u)\partial_t u = 0,$$

118   *Local Positivity of Intersections*

where $Q := i - J \in C^\infty(\mathbb{C}^n, \mathrm{End}_\mathbb{R}(\mathbb{C}^n))$. The rescaled map $\hat{u}$ then satisfies

$$\bar{\partial}\hat{u} - \hat{Q}(\hat{u})\partial_t\hat{u} = 0 \tag{B.7}$$

if we define $\hat{J}, \hat{Q}\colon \mathbb{C}^n \to \mathrm{End}_\mathbb{R}(\mathbb{C}^n)$ by

$$\hat{J}(p) := J(p/R), \qquad \hat{Q} := i - \hat{J}.$$

The advantage of these definitions is that $\hat{Q}$ can be made arbitrarily C^∞-small by choosing $R > 0$ large, and since $u\colon \mathbb{D} \to \mathbb{C}^n$ is continuous with $u(0) = 0$, one can subsequently choose $\rho > 0$ small to make the function $\hat{Q} \circ \hat{u}\colon \mathbb{D} \to \mathrm{End}_\mathbb{R}(\mathbb{C}^n)$ correspondingly small so that (B.7) becomes a small perturbation of the linear equation $\bar{\partial}\hat{u} = 0$. In this context, it will be useful to note that for any $k \in \mathbb{N}$ and $p \in (1, \infty)$ with $kp > 2$, there exist constants $c > 0$ and $\gamma > 0$ such that every $f \in W^{k,p}(\mathbb{D})$ with $f(0) = 0$ is related to its rescaled cousin $\hat{f}(z) := Rf(\rho z)$ by

$$\|\hat{f}\|_{W^{k,p}(\mathbb{D})} \le cR\rho^\gamma\|f\|_{W^{k,p}(\mathbb{D})}. \tag{B.8}$$

This can be proved as a corollary of the Sobolev embedding theorem, and it implies that, for each $k \ge 2$, $\|\hat{u}\|_{H^k}$ can also be made arbitrarily small by choosing $\rho > 0$ small for any given $R > 0$. In the following, we always reserve the right to enlarge R and subsequently shrink ρ whenever convenient.

Arguing by induction, the goal is now to show that if $\hat{u} \in H^k(\mathbb{D})$ for a given $k \in \mathbb{N}$, then $\hat{u}$ is also of class H^{k+1} on $\mathbb{D}_r$ for some $r < 1$, where in the case $k = 1$ we impose the extra hypothesis $\hat{u} \in W^{1,\infty}(\mathbb{D})$. As in Proposition B.13, the argument uses difference quotients; e.g., if one can prove uniform H^k-bounds on the difference quotients $D^h_s\hat{u}$ with respect to s as $h \to 0$, then the Banach–Alaoglu theorem implies that $\partial_s\hat{u}$ is in H^k. The assumption $\hat{u} \in H^k(\mathbb{D})$ already implies a uniform H^{k-1}-bound on $D^h_s\hat{u}$ as $h \to 0$, where in the case $k = 1$, there is an additional L^∞-bound. Choosing a smooth bump function $\beta\colon \mathbb{D} \to [0, 1]$ with compact support in the interior and $\beta|_{\mathbb{D}_r} \equiv 1$ for some $r < 1$, it then suffices to find a uniform bound on $\|\beta D^h_s\hat{u}\|_{H^k}$ as $h \to 0$. The usual estimate (B.5) gives

$$\|\beta D^h_s\hat{u}\|_{H^k} \le c\|\bar{\partial}(\beta D^h_s\hat{u})\|_{H^{k-1}}.$$

To bound the right-hand side, one can apply the operator D^h_s to the equation $\bar{\partial}\hat{u} = (\hat{Q} \circ \hat{u})\partial_t\hat{u}$ from (B.7), giving

$$\bar{\partial}(D^h_s\hat{u}) = D^h_s(\hat{Q} \circ \hat{u})\partial_t\hat{u} + (\hat{Q} \circ \hat{u})\partial_t(D^h_s\hat{u})$$

and thus

$$\bar{\partial}(\beta D^h_s\hat{u}) = \beta D^h_s(\hat{Q} \circ \hat{u})\partial_t\hat{u} + (\hat{Q} \circ \hat{u})\partial_t(\beta D^h_s\hat{u}) + \left(\bar{\partial}\beta - (\hat{Q} \circ \hat{u})\partial_t\beta\right)D^h_s\hat{u}.$$

B.1 Regularity and the Similarity Principle

From this we deduce the estimate

$$\|\beta D_s^h \hat{u}\|_{H^k} \leq c \left\|\hat{Q}(\hat{u})\partial_t(\beta D_s^h \hat{u})\right\|_{H^{k-1}} + c \left\|\beta D_s^h(\hat{Q}\circ\hat{u})\partial_t\hat{u}\right\|_{H^{k-1}}$$
$$+ c \left\|\left(\bar{\partial}\beta - \hat{Q}(\hat{u})\partial_t\beta\right)D_s^h u\right\|_{H^{k-1}}. \tag{B.9}$$

We claim that after suitable adjustments of the rescaling parameters ρ and R, every term in (B.9) is bounded uniformly as $h \to 0$.

Indeed, since $\hat{Q}$ can be assumed arbitrarily C^k-small on the image of $\hat{u}$, we can also apply (B.8) if $k \geq 2$ to assume that the composition $\hat{Q}\circ\hat{u}$ is arbitrarily small in H^k, in which case the continuous product pairing $H^k \times H^{k-1} \to H^{k-1}$ gives a uniform bound on the third term. This argument does not quite work in the case $k = 1$, as (B.8) is then not valid and there is no continuous product pairing $H^1 \times L^2 \to L^2$, but here one can instead make $\hat{Q}\circ\hat{u}$ arbitrarily C^0-small and achieve a uniform L^2-bound.

For the first term, if $k \geq 2$, then one can similarly use the continuous product pairing $H^k \times H^{k-1} \to H^{k-1}$ and make $\|\hat{Q}\circ\hat{u}\|_{H^k}$ small via (B.8), giving an estimate of the form

$$\left\|\hat{Q}(\hat{u})\partial_t\left(\beta D_s^h \hat{u}\right)\right\|_{H^{k-1}} \leq \delta \left\|\partial_t\left(\beta D_s^h \hat{u}\right)\right\|_{H^{k-1}} \leq c\delta\|\beta D_s^h \hat{u}\|_{H^k},$$

with a constant $\delta > 0$ that can be made arbitrarily small by suitable adjustments of R and ρ. One can therefore absorb this term into the left-hand side of (B.9). Once again, a special argument is required for the case $k = 1$, but here one can instead assume $\hat{Q}\circ\hat{u}$ is C^0-small and use the continuous pairing $C^0 \times L^2 \to L^2$ to achieve the same result.

The second term in (B.9) requires some version of the chain rule for the difference quotient operator D_s^h. Here one can write

$$\hat{Q}(p + p') = \hat{Q}(p) + \int_0^1 d\hat{Q}(p + \tau p')p' \, d\tau$$
$$= \hat{Q}(p) + d\hat{Q}(p)p' + \left(\int_0^1 \left[d\hat{Q}(p + \tau p') - d\hat{Q}(p)\right] d\tau\right)p'$$
$$= \hat{Q}(p) + \left[d\hat{Q}(p) + \hat{R}(p, p')\right]p'$$

for a smooth remainder function $\hat{R}\colon \mathbb{C}^n\times\mathbb{C}^n \to \mathrm{Hom}_{\mathbb{R}}(\mathbb{C}^n, \mathrm{End}_{\mathbb{R}}(\mathbb{C}^n))$ satisfying $\hat{R}(\cdot, 0) \equiv 0$, and use this to derive a formula of the form

$$D_s^h(\hat{Q}\circ\hat{u}) = \left[d\hat{Q}\circ\hat{u} + \hat{R}\circ\left(\hat{u}, hD_s^h\hat{u}\right)\right]D_s^h\hat{u}, \tag{B.10}$$

valid for all $h \neq 0$ sufficiently close to 0. For $k \geq 2$, one can use (B.8) to assume the terms $\hat{u}$ and $hD_s^h\hat{u}$ satisfy an arbitrarily small H^k-bound independent of h, and then use the smoothness of $\hat{Q}$ and $\hat{R}$ and the fact that $\hat{R}(\cdot, 0) \equiv 0$ to assume

120   *Local Positivity of Intersections*

that the bracketed term in the above expression is arbitrarily H^k-small for all h near 0. Since H^k is a Banach algebra, this gives rise to an estimate of the form

$$\left\| \beta D_s^h \left(\widehat{Q} \circ \widehat{u} \right) \right\|_{H^k} \le \delta \| \beta D_s^h \widehat{u} \|_{H^k},$$

where the constant $\delta > 0$ can be assumed arbitrarily small after adjusting R and ρ. Since $\| \partial_t \widehat{u} \|_{H^{k-1}} \le \| \widehat{u} \|_{H^k}$ can also be assumed small by (B.8) and the product pairing $H^k \times H^{k-1} \to H^{k-1}$ is continuous, it follows that the second term in (B.9) can also be absorbed into the left-hand side. In the case $k = 1$, we can instead use the uniform L^∞-bound on $\partial_t u$ to put a bound on $\| \partial_t \widehat{u} \|_{L^\infty}$ while making ρ as small as is needed, and then use the uniform L^2-bound on $D_s^h \widehat{u}$ to derive from (B.10) a uniform L^2-bound on $D_s^h(\widehat{Q} \circ \widehat{u})$, which now directly implies a uniform bound on the second term in (B.9).   □

Exercise B.16   Show that, for any $\varphi \in C_0^\infty(\mathbb{D})$ and $f \in C^0(\mathbb{D})$ such that $f|_{\mathbb{D} \setminus \{0\}}$ is of class C^1 with bounded derivative, the usual formula for integration by parts

$$\int_{\mathbb{D}} \partial_j f \cdot \varphi = - \int_{\mathbb{D}} f \cdot \partial_j \varphi$$

is valid, and deduce that f belongs to $W^{1,\infty}(\mathbb{D})$.

B.1.3 Local Existence of Holomorphic Sections

The main engine behind the similarity principle is the following local existence result for solutions to linear Cauchy–Riemann type equations.

Theorem B.17   *Assume* $2 < p \le \infty$ *and* $A \in L^p(\mathbb{D}, \mathrm{End}_\mathbb{R}(\mathbb{C}^n))$. *Then, for sufficiently small* $\rho > 0$, *the problem*

$$\bar{\partial} u + A u = 0$$

$$u(0) = u_0$$

admits a weak solution $u \in C^0(\mathbb{D}_\rho) \cap H^1(\mathbb{D}_\rho)$ *for every* $u_0 \in \mathbb{C}^n$.

Notice that by elliptic regularity (Proposition B.13), the local solutions $u \colon \mathbb{D}_\rho \to \mathbb{C}^n$ provided by this theorem may be much nicer than just continuous functions with weak derivatives in L^2; e.g., they will be smooth if A is smooth. One easy consequence is the following fundamental result in complex geometry, which gives an equivalence between smooth complex-linear Cauchy–Riemann type operators and holomorphic vector bundle structures.

Corollary B.18   *Suppose* E *is a complex vector bundle over a Riemann surface* Σ, *and* $\mathbf{D} \colon \Gamma(E) \to \Gamma(\overline{\mathrm{Hom}}_\mathbb{C}(T\Sigma, E))$ *is a smooth complex-linear*

B.1 Regularity and the Similarity Principle 121

Cauchy–Riemann type operator. Then E admits a unique maximal atlas of smooth local complex trivializations whose transition maps are holomorphic such that a section $\eta \in \Gamma(E|_{\mathcal{U}})$ defined on some open domain $\mathcal{U} \subset \Sigma$ is holomorphic with respect to these trivializations if and only if $\mathbf{D}\eta = 0$.

Proof For any point $p \in \Sigma$, Theorem B.17 and Proposition B.13 together provide a collection of smooth sections $\eta_1, \ldots, \eta_n$ defined on a neighborhood of p that all satisfy $\mathbf{D}\eta_i = 0$ and are pointwise complex-linearly independent at p (and therefore also in a neighborhood of p). We define the desired atlas of local trivializations by viewing collections of this sort as local frames. The Leibniz rule for complex-linear Cauchy–Riemann type operators (cf. Remark B.2) then implies that transition maps are holomorphic. $\square$

The local existence theorem admits a fairly straightforward proof using the $p > 2$ case of Proposition B.10. The idea is to multiply A by the characteristic function χ_ρ of $\mathbb{D}_\rho$ for $\rho > 0$, producing a family of bounded linear operators

$$\mathbf{D}_\rho := \bar{\partial} + \chi_\rho A : W^{1,p}(\mathbb{D}) \to L^p(\mathbb{D}),$$

which converge in the norm topology to $\bar{\partial} : W^{1,p}(\mathbb{D}) \to L^p(\mathbb{D})$ as $\rho \to 0$. It follows that the operators

$$\mathbf{L}_\rho : W^{1,p}(\mathbb{D}) \to L^p(\mathbb{D}) \times \mathbb{C}^n : u \mapsto (\mathbf{D}_\rho u, u(0))$$

also converge as $\rho \to 0$ to $\mathbf{L}_0(u) = (\bar{\partial}u, u(0))$; note here that $W^{1,p}(\mathbb{D}) \to \mathbb{C}^n : u \mapsto u(0)$ is a well-defined and continuous linear map due to the Sobolev embedding theorem. Since $\bar{\partial} : W^{1,p}(\mathbb{D}) \to L^p(\mathbb{D})$ has a bounded right inverse and holomorphic functions on $\mathbb{D}$ can take arbitrary values at a point, the operator $\mathbf{L}_0$ also has a bounded right inverse, and so therefore does $\mathbf{L}_\rho$ for $\rho > 0$ sufficiently small, as the existence of bounded right inverses is an open condition. The right inverse of $\mathbf{L}_\rho$ can then be used to produce functions $u \in W^{1,p}(\mathbb{D})$ that have prescribed values at 0 and satisfy $(\bar{\partial} + \chi_\rho A)u = 0$, so in particular they satisfy $(\bar{\partial} + A)u = 0$ on $\mathbb{D}_\rho$. These functions are also continuous since, by the Sobolev embedding theorem, $W^{1,p}(\mathbb{D})$ embeds continuously into $C^0(\mathbb{D})$ for $p > 2$.

The argument just sketched would not work for $p = 2$ because $H^1(\mathbb{D}) = W^{1,2}(\mathbb{D})$ does not embed into $C^0(\mathbb{D})$. Since we did not prove the $p > 2$ case of Proposition B.10, we will have to do something slightly more roundabout in order to produce a self-contained proof of local existence. It is based on the following lemma, which was suggested by Jean-Claude Sikorav:

Lemma B.19 *Under the same assumptions as in Theorem B.17, suppose $0 < r \leq 1$ and $f_0 : \mathbb{D}_r \to \mathbb{C}^n$ is a holomorphic function. Then for any $\delta > 0$, there*

122 *Local Positivity of Intersections*

exists $\rho \in (0, r]$ and a continuous function $f: \mathbb{D}_\rho \to \mathbb{C}^n$ such that $|f| \leq \delta$ and $u := f_0 + f: \mathbb{D}_\rho \to \mathbb{C}^n$ is a weak solution to the equation $(\bar{\partial} + A)u = 0$.

Proof We will look for a continuous weak solution $u: \mathbb{D} \to \mathbb{C}^n$ to the equation

$$(\bar{\partial} + A_\rho)u = 0,$$

for some small number $\rho > 0$, where $A_\rho := \chi_\rho A$ and $\chi_\rho: \mathbb{D} \to [0, 1]$ denotes the function that equals 1 on $\mathbb{D}_\rho$ and 0 everywhere else. We claim that each of the operators $\bar{\partial} + A_\rho$ is a bounded linear operator $H^1(\mathbb{D}) \to L^2(\mathbb{D})$ and that these operators converge in the operator norm to $\bar{\partial}$ as $\rho \to 0$. Recall that $H^1(\mathbb{D})$ is a "Sobolev borderline case," so it admits continuous inclusions $H^1(\mathbb{D}) \hookrightarrow L^q(\mathbb{D})$ for every finite $q \geq 1$ (see [AF03]). Thus if we pick $q > 1$ according to the condition $1/q + 2/p = 1$, then Hölder's inequality and the continuous inclusion $H^1 \hookrightarrow L^{2q}$ imply that for any $u \in H^1(\mathbb{D})$,

$$\|A_\rho u\|_{L^2}^2 \leq \int_{\mathbb{D}_\rho} |A|^2 |u|^2 \leq \left\||A|^2\right\|_{L^{p/2}(\mathbb{D}_\rho)} \cdot \left\||u|^2\right\|_{L^q(\mathbb{D}_\rho)} \leq \|A\|_{L^p(\mathbb{D}_\rho)}^2 \|u\|_{L^{2q}(\mathbb{D})}^2$$
$$\leq c\|A\|_{L^p(\mathbb{D}_\rho)}^2 \|u\|_{H^1(\mathbb{D})}^2$$

for some constant $c > 0$. This proves the claim, since $\|A\|_{L^p(\mathbb{D}_\rho)} \to 0$ as $\rho \to 0$.

Since $\bar{\partial}: H^1(\mathbb{D}) \to L^2(\mathbb{D})$ has a bounded right inverse $T: L^2(\mathbb{D}) \to H^1(\mathbb{D})$, it follows that $\bar{\partial} + A_\rho$ also has a bounded right inverse

$$T_\rho: L^2(\mathbb{D}) \to H^1(\mathbb{D})$$

for all $\rho > 0$ sufficiently small. It should now at least seem plausible that any solution $f_0 \in H^1(\mathbb{D})$ to $\bar{\partial}f_0 = 0$ admits an H^1-close perturbation $f_0 + f$ satisfying $(\bar{\partial} + A_\rho)(f_0 + f) = 0$; indeed, the latter is equivalent to the equation

$$(\bar{\partial} + A_\rho)f = -A_\rho f_0,$$

which can be solved by

$$f := -T_\rho(A_\rho f_0).$$

This function clearly is H^1-small whenever ρ is correspondingly small since $T_\rho: L^2 \to H^1$ is continuous and $L^p(\mathbb{D})$ embeds continuously into $L^2(\mathbb{D})$; hence

$$\|A_\rho f_0\|_{L^2(\mathbb{D})} \leq c\|A_\rho f_0\|_{L^p(\mathbb{D})} \leq c\|A\|_{L^p(\mathbb{D}_\rho)}\|f_0\|_{C^0} \to 0 \quad \text{as} \quad \rho \to 0. \quad (\text{B.11})$$

We claim, in fact, that the operator T_ρ can be chosen to make f continuous and C^0-small when ρ is correspondingly small. By (B.11), this will be immediate if we can show that T_ρ restricts to a continuous linear map $L^p(\mathbb{D}) \to C^0(\mathbb{D})$ for $p > 2$, a fact that we already know is true of $T: L^2(\mathbb{D}) \to H^1(\mathbb{D})$ by

B.1 Regularity and the Similarity Principle 123

Proposition B.12. Thus to prove the claim, let us write down a more explicit definition of T_ρ. Notice that

$$(\bar{\partial} + A_\rho)T = 1 + A_\rho T$$

is a bounded linear operator on L^2 and is close to the identity in the operator norm since $T : L^2 \to H^1$ is continuous and $A_\rho : H^1 \to L^2$ is small. But, for slightly different reasons, this operator is also close to the identity in the space of bounded linear operators on L^p; indeed, $A_\rho : C^0 \to L^p$ is also continuous and small since

$$\|A_\rho u\|_{L^p(\mathbb{D})} \leq \|A_\rho\|_{L^p(\mathbb{D})}\|u\|_{C^0(\mathbb{D})} = \|A\|_{L^p(\mathbb{D}_\rho)}\|u\|_{C^0(\mathbb{D})},$$

so this statement follows from the continuity of $T : L^p(\mathbb{D}) \to C^0(\mathbb{D})$. Thus for any $\rho > 0$ sufficiently small, $1 + A_\rho T$ defines isomorphisms on both $L^2(\mathbb{D})$ and $L^p(\mathbb{D})$, so that defining

$$T_\rho := T(1 + A_\rho T)^{-1}$$

gives a right inverse of $\bar{\partial} + A_\rho$ that is continuous both from L^2 to H^1 and from L^p to C^0. $\qquad\square$

Proof of Theorem B.17 Using Lemma B.19, we can construct the columns of a continuous matrix-valued function $\Phi : \mathbb{D}_\rho \to \text{End}_\mathbb{R}(\mathbb{C}^n)$ for $\rho > 0$ small such that Φ weakly satisfies $(\bar{\partial} + A)\Phi = 0$ and is arbitrarily C^0-close to the constant (and thus holomorphic) function $\Phi_0(z) := 1$. We can therefore assume Φ takes values in $\text{GL}(2n, \mathbb{R})$. Continuous solutions $u : \mathbb{D}_\rho \to \mathbb{C}^n$ to $(\bar{\partial}+A)u = 0$ with prescribed values $u(0) = u_0$ can then be constructed by multiplying Φ by suitable constant vectors in $\mathbb{C}^n$. $\qquad\square$

B.1.4 The Similarity Principle

We can now prove the main result of the present section.

Theorem B.20 *Assume E is a complex vector bundle of class C^1 over a Riemann surface Σ, $\mathbf{D}$ is a linear Cauchy–Riemann type operator on E of class L^p for some $p \in (2, \infty]$ in the sense of Remark B.6, and $\eta : \Sigma \to E$ is a continuous section that is a weak solution to the equation $\mathbf{D}\eta = 0$ with $\eta(z_0) = 0$ for some point $z_0 \in \Sigma$. Then there exists a continuous local complex trivialization of E near z_0 that identifies η with a holomorphic function. Moreover, if $\mathbf{D}$ is smooth and complex linear, then the local trivialization near z_0 can be arranged to be smooth.*

124 *Local Positivity of Intersections*

Proof The issue is purely local, so assume $A \in L^p(\mathbb{D}, \mathrm{End}_{\mathbb{R}}(\mathbb{C}^n))$ with $p > 2$ and $u\colon \mathbb{D} \to \mathbb{C}^n$ is a continuous weak solution to

$$(\bar{\partial} + A)u = 0$$

with $u(0) = 0$. We start by replacing $\bar{\partial} + A$ by another Cauchy–Riemann type operator that is complex linear but has the same regularity. Indeed, choose a measurable function $C\colon \mathbb{D} \to \mathrm{End}_{\mathbb{C}}(\mathbb{C}^n)$ such that $|C(z)| \leq |A(z)|$ and $C(z)u(z) = A(z)u(z)$ for all $z \in \mathbb{D}$. Then C is also of class $L^p(\mathbb{D})$ and u also satisfies $\bar{\partial}u + Cu = 0$.

Now construct a local frame as in the proof of Corollary B.18; that is, let $\Phi\colon \mathbb{D}_\rho \to \mathrm{End}_{\mathbb{C}}(\mathbb{C}^n)$ be a complex matrix-valued function whose columns are local weak solutions to $(\bar{\partial} + C)\eta = 0$ as provided by Theorem B.17, with $\Phi(0) = \mathbb{1}$. Since $C \in L^p(\mathbb{D})$ with $p > 2$, Φ is in $C^0(\mathbb{D}_\rho) \cap H^1(\mathbb{D}_\rho)$, and it also satisfies $(\bar{\partial} + C)\Phi = 0$. After shrinking $\rho > 0$ if necessary, continuity then implies that we are free to assume $\Phi(z)$ is invertible for all $z \in \mathbb{D}_\rho$, and we can therefore define a continuous function $f\colon \mathbb{D}_\rho \to \mathbb{C}^n$ by

$$f(z) := [\Phi(z)]^{-1}u(z).$$

To conclude, we need to show that f is a weak solution to $\bar{\partial}f = 0$, in which case Proposition B.13 implies that f is also smooth, and therefore holomorphic. If Φ, u and f were all smooth, then $\bar{\partial}f = 0$ would follow from the fact that $\bar{\partial} + C$ is complex linear and annihilates both Φ and u, as the Leibniz rule (cf. Remark B.2) then implies

$$0 = (\bar{\partial} + C)u = (\bar{\partial} + C)(\Phi f) = \left[(\bar{\partial} + C)\Phi\right]f + \Phi(\bar{\partial}f) = \Phi(\bar{\partial}f).$$

An additional argument is required in order to justify this conclusion without knowing whether Φ and u are smooth. What we do know is that u is continuous and Φ is in both C^0 and H^1; since $A \in L^p(\mathbb{D}) \subset L^2(\mathbb{D})$, we also have $-Au \in L^2(\mathbb{D})$, so that Proposition B.13 implies that u is also in $H^1(\mathbb{D}_\rho)$. To make use of this, we can consider the following normed linear space:

$$X := H^1(\mathbb{D}_\rho) \cap C^0(\mathbb{D}_\rho), \qquad \|\eta\|_X := \|\eta\|_{H^1(\mathbb{D}_\rho)} + \|\eta\|_{C^0(\mathbb{D}_\rho)}.$$

It is a straightforward exercise to prove that X has the following properties:

- X is complete; i.e., it is a Banach space.
- $C^\infty(\mathbb{D}_\rho) \cap X$ is dense in X. (Indeed, one can check that the standard mollification procedure for functions in $H^1(\mathbb{D}_\rho)$ as in [Eva98, §5.3] works simultaneously for $C^0(\mathbb{D}_\rho)$).
- X is a Banach algebra; i.e., there exists a continuous product pairing $X \times X \to X\colon (g, h) \mapsto gh$ for complex-valued functions, and there is similarly a

B.1 Regularity and the Similarity Principle

continuous product pairing $X \times L^2(\mathbb{D}_\rho) \to L^2(\mathbb{D}_\rho)$. (The main tool in both cases is the inequality $\|gh\|_{L^2} \le \|g\|_{C^0}\|h\|_{L^2}$ for $g \in C^0$ and $h \in L^2$.)

- If $\Phi \in X$ is a function $\mathbb{D} \to \text{End}_{\mathbb{C}}(\mathbb{C}^n)$ with image in $\text{GL}(n, \mathbb{C})$, then the function $\Phi^{-1}(z) := [\Phi(z)]^{-1}$ also belongs to X and depends continuously on $\Phi \in X$ in the topology of X. (Recall that $\text{GL}(n, \mathbb{C}) \to \text{GL}(n, \mathbb{C})$: $B \mapsto B^{-1}$ is a smooth function.)

Notice that, for any $g \in X$, we have $\bar{\partial}g \in L^2(\mathbb{D}_\rho)$ since $g \in H^1(\mathbb{D}_\rho)$, and, similarly, $Cg \in L^2(\mathbb{D}_\rho)$ since $C \in L^p \subset L^2$ and g is continuous. In fact, $\bar{\partial} + C$ defines a continuous linear operator

$$(\bar{\partial} + C): X \to L^2(\mathbb{D}_\rho).$$

The previous remarks now imply after taking $\rho > 0$ sufficiently small that Φ^{-1} and u both belong to X, so, by the Banach algebra property, so does $f = \Phi^{-1}u$. Now use the density of smooth functions to find sequences of smooth functions f_ν, Φ_ν converging in X to f and Φ, respectively, so that (using the Banach algebra property again) $u_\nu := \Phi_\nu f_\nu$ also converges in X to u. The Leibniz rule now gives

$$(\bar{\partial} + C)u_\nu = \left[(\bar{\partial} + C)\Phi_\nu\right]f_\nu + \Phi_\nu(\bar{\partial}f_\nu),$$

in which the left-hand side and the first term on the right-hand side both converge in L^2 as $\nu \to \infty$ to zero, while the last term converges in L^2 to $\Phi(\bar{\partial}f)$, proving $\bar{\partial}f = 0$.

Finally, consider the special case in which A is smooth and complex linear. Under this assumption, Corollary B.18 implies that $\mathbf{D}$ defines a holomorphic structure on E in which η is a holomorphic section. Alternatively, one could instead apply the argument above after setting $C := A$ in the initial step, so that the function $\Phi: \mathbb{D}_\rho \to \text{End}_{\mathbb{C}}(\mathbb{C}^n)$ satisfying $(\bar{\partial} + C)\Phi$ is then smooth by elliptic regularity (Corollary B.14). $\qquad\square$

Corollary B.21 *Under the assumptions of Theorem B.20, suppose η is not identically zero near z_0, and choose local holomorphic coordinates and a local complex trivialization near z_0 to identify η with a function $\mathbb{D} \to \mathbb{C}^n$ such that $z_0 = 0 \in \mathbb{D}$. Then η satisfies the formula*

$$\eta(z) = z^k C + |z|^k R(z)$$

for some $k \in \mathbb{N}$, $C \in \mathbb{C}^n \setminus \{0\}$ and a function $R(z) \in \mathbb{C}^n$ such that $\lim_{z \to 0} R(z) = 0$.

Proof In the chosen coordinates and trivialization, the similarity principle provides a continuous transition map $\Phi: \mathbb{D}_\rho \to \text{GL}(n, \mathbb{C})$ for $\rho > 0$ small and a holomorphic function $f: \mathbb{D}_\rho \to \mathbb{C}^n$ such that $\eta = \Phi f$ and $f(0) = 0$. Since η

126   *Local Positivity of Intersections*

is not identically zero on this neighborhood, we have $f(z) = z^k g(z)$ for some $k \in \mathbb{N}$ and a holomorphic function $g \colon \mathbb{D}_\rho \to \mathbb{C}^n$ with $g(0) \neq 0$. Then

$$u(z) = z^k \Phi(0)g(0) + z^k \left[\Phi(z)g(z) - \Phi(0)g(0)\right],$$

in which $\Phi(0)g(0) \neq 0$ and the term in brackets is a continuous function that vanishes at $z = 0$. $\qquad\square$

Remark B.22 If η in the corollary above is smooth, then the result is equivalent to the statement that the Taylor series of η about z_0 is nontrivial and its lowest-order term is holomorphic (i.e., a polynomial in z with no dependence on $\bar{z}$). However, the result remains valid even if E, $\mathbf{D}$ and η are not assumed smooth; e.g., in the proof of the representation formula in the next section, we will need to consider examples where η is only known to be of class C^1.

B.2 The Representation Formula

If $u(z) = v(\zeta)$ is an isolated intersection of two J-holomorphic curves in an almost complex 4-manifold and at least one of the curves is immersed at the intersection point, then there is a relatively easy argument via the similarity principle (see [MS12, Exercise 2.6.1]) to prove that this intersection must count positively. The same holds without assuming that either curve is immersed, but the proof requires more work. One approach, due to McDuff [McD94], shows that a J-holomorphic curve with critical points always admits a global perturbation to an *immersed J'-*holomorphic curve for some perturbed almost complex structure J', and thus the general case can be reduced to the immersed case. This is an elegant argument, but it gives little insight as to what is really happening near critical points of holomorphic curves, so we will instead discuss a purely local approach, using a variation on a result of Micallef and White [MW95]. The following statement is weaker than the actual Micallef–White theorem but suffices for our purposes, and is easier to prove.

Theorem B.23 *Suppose (M, J) is a smooth almost complex manifold of dimension $2n$, and $u \colon (\Sigma, j) \to (M, J)$ is a J-holomorphic curve that is not constant in some neighborhood of the point $z_0 \in \Sigma$. Then there exists a unique integer $k \in \mathbb{N}$ and 1-dimensional complex subspace $L \subset T_{u(z_0)}M$ such that one can find a C^∞-smooth coordinate chart on a neighborhood of $u(z_0) \in M$ and a C^1-smooth coordinate chart on a neighborhood of $z_0 \in \Sigma$, identifying these points with the origin in $\mathbb{C}^n$ and $\mathbb{C}$, respectively, and identifying L with $\mathbb{C} \times \{0\} \subset \mathbb{C}^n$, so that u in these coordinates near z_0 takes the form*

$$u(z) = (z^k, \widehat{u}(z)) \in \mathbb{C} \times \mathbb{C}^{n-1}$$

B.2 The Representation Formula

for some C^1-smooth function $\widehat{u}(z) \in \mathbb{C}^{n-1}$ defined near $z = 0$ and satisfying $\widehat{u}(z) = O(|z|^{k+1})$. Moreover, the C^1-smooth chart near z_0 may be assumed C^∞ at all points other than z_0, and $\widehat{u}$ either vanishes identically or satisfies the formula

$$\widehat{u}(z) = z^{k+\ell_u} C_u + |z|^{k+\ell_u} r_u(z)$$

for some constants $C_u \in \mathbb{C}^{n-1} \setminus \{0\}$, $\ell_u \in \mathbb{N}$, and a function $r_u(z) \in \mathbb{C}^{n-1}$ with $r_u(z) \to 0$ as $z \to 0$. We will say in this situation that u has **tangent space** L with **critical order** $k - 1$ at z_0.

Further, if $v \colon (\Sigma', j') \to (M, J)$ is another nonconstant J-holomorphic curve with an intersection $u(z_0) = v(\zeta_0)$ at some point $\zeta_0 \in \Sigma'$ where u and v have the same tangent spaces and critical orders, then the coordinates above can be chosen together with C^1-smooth coordinates near $\zeta_0 \in \Sigma'$ having the same properties, in particular such that v satisfies a representation formula

$$v(z) = (z^k, \widehat{v}(z)),$$

with either $\widehat{v} \equiv 0$ or

$$\widehat{v}(z) = z^{k+\ell_v} C_v + |z|^{k+\ell_v} r_v(z)$$

for some $C_v \in \mathbb{C}^{n-1} \setminus \{0\}$, $\ell_v \in \mathbb{N}$ and function $r_v(z)$ with $r_v(z) \to 0$ as $z \to 0$.

Finally, any two curves written in this way are related to each other as follows: either $\widehat{u} \equiv \widehat{v}$, or

$$\widehat{v}(z) - \widehat{u}(z) = z^{k+\ell'} C' + |z|^{k+\ell'} r'(z), \tag{B.12}$$

for some constants $C' \in \mathbb{C}^{n-1} \setminus \{0\}$, $\ell' \in \mathbb{N}$ and a function $r'(z) \in \mathbb{C}^{n-1}$ with $r'(z) \to 0$ as $z \to 0$.

Exercise B.24 Prove Theorem B.23 for the case $(M, J) = (\mathbb{C}^n, i)$.

Exercise B.25 Use Theorem B.23 to show that for any J-holomorphic curve $u \colon (\Sigma, j) \to (M, J)$ with a point $z_0 \in \Sigma$ where $du(z_0) = 0$ but u is not constant near z_0, all other points in some neighborhood of z_0 are immersed points, and, moreover, the natural map

$$z \mapsto \operatorname{im} du(z)$$

from the immersed points in Σ to the bundle of complex 1-dimensional subspaces in (TM, J) extends continuously to z_0.

The much deeper theorem of Micallef and White [MW95] applies to a more general class of maps than just J-holomorphic curves, and it also provides coordinates in which $\widehat{u}(z)$ and $\widehat{v}(z)$ become *polynomials*, and thus the remainder formulas stated in Theorem B.23 become obvious. The Micallef–White

128 *Local Positivity of Intersections*

theorem is discussed in more detail in [MS12, Appendix E] (written with Laurent Lazzarini) and [Sik97]. Our weaker version is based on ideas due to Hofer, and is essentially a "nonasymptotic version" of Siefring's relative asymptotic analysis [Sie05] described in Lecture 3.

The remainder of §B.2 will be concerned with the proof of Theorem B.23.

B.2.1 The Generalized Tangent-Normal Decomposition

The first step is to prove a refined version of the corollary that was observed in Exercise B.25.

Proposition B.26 *If $u: (\Sigma, j) \to (M, J)$ is a smooth connected J-holomorphic curve that is not constant, then the critical points of u are isolated, and there exists a unique smooth rank 1 complex subbundle*

$$T_u \subset u^*TM$$

such that $(T_u)_z = \operatorname{im} du(z)$ for all immersed points $z \in \Sigma$ of u. Moreover, du defines a smooth section of the complex line bundle $\operatorname{Hom}_{\mathbb{C}}(T\Sigma, T_u)$ whose zeroes coincide with the critical points of z, and these zeroes all have positive order.

We shall refer to the subbundle $T_u \subset u^*TM$ in Proposition B.26 as the **generalized tangent bundle** of the curve $u: (\Sigma, j) \to (M, J)$ and define the **critical order** of each critical point of u to be the order of the corresponding zero of $du \in \Gamma(\operatorname{Hom}_{\mathbb{C}}(T\Sigma, T_u))$. A choice of smooth complex subbundle $N_u \subset u^*TM$ that is complementary to T_u will then be referred to as the **generalized normal bundle** of u, characterized by the smooth complex-linear splitting

$$u^*TM = T_u \oplus N_u.$$

The bundle N_u is nonunique but is clearly unique up to isomorphism, so we shall typically ignore this detail in our discussion – if you prefer, you are free to eliminate the ambiguity by assuming always that N_u is the orthogonal complement of T_u with respect to some fixed choice of J-invariant Riemannian metric.

Proposition B.26 is an easy consequence of the correspondence given by Corollary B.18 between complex-linear Cauchy–Riemann operators and holomorphic bundle structures. It depends on the following trick borrowed from [IS99]. Consider the linearized Cauchy–Riemann operator

$$\mathbf{D}_u: \Gamma(u^*TM) \to \Gamma(\overline{\operatorname{Hom}}_{\mathbb{C}}(T\Sigma, u^*TM)),$$

B.2 The Representation Formula   129

which can be defined via the property that if $\{u_\sigma : \Sigma \to M\}_{\sigma\in(-\epsilon,\epsilon)}$ is any smooth 1-parameter family of maps satisfying $u_0 = u$ and $\partial_\sigma u_\sigma|_{\sigma=0} = \eta \in \Gamma(u^*TM)$, then for any connection ∇ on M and any $z \in \Sigma$ and $X \in T_z\Sigma$,

$$(\mathbf{D}_u\eta)(X) = \nabla_\sigma\left[(\bar\partial_J u_\sigma)(X)\right]\Big|_{\sigma=0},$$

where $\bar\partial_J$ denotes the nonlinear Cauchy–Riemann operator

$$\bar\partial_J f := df + J \circ df \circ j \in \Gamma(\overline{\mathrm{Hom}}_{\mathbb{C}}(T\Sigma, f^*TM)).$$

Choosing the connection ∇ to be symmetric, one can derive a more direct formula for $\mathbf{D}_u$ in the form

$$\mathbf{D}_u\eta = \nabla\eta + J(u) \circ \nabla\eta \circ j + (\nabla_\eta J) \circ Tu \circ j,$$

which shows that $\mathbf{D}_u$ is indeed a smooth linear Cauchy–Riemann type operator. In general, $\mathbf{D}_u$ is real but not complex linear, because the connection ∇ need not be complex and $\nabla_{J\eta}J - J\nabla_\eta J$ need not vanish. On the other hand, it is easy to check that the complex-linear part of $\mathbf{D}_u$,

$$\mathbf{D}_u^{\mathbb{C}} : \Gamma(u^*TM) \to \Gamma(\overline{\mathrm{Hom}}_{\mathbb{C}}(T\Sigma, u^*TM))$$

$$\eta \mapsto \frac{1}{2}\left(\mathbf{D}_u\eta - J\mathbf{D}_u(J\eta)\right),$$

also satisfies the required Leibniz rule and is thus a smooth complex-linear Cauchy–Riemann type operator. By Corollary B.18, $\mathbf{D}_u^{\mathbb{C}}$ therefore determines a holomorphic vector bundle structure on u^*TM.

Lemma B.27 *The complex-linear bundle map $du: T\Sigma \to u^*TM$ is holomorphic with respect to the canonical holomorphic structure of $T\Sigma$ and the holomorphic structure on u^*TM determined by $\mathbf{D}_u^{\mathbb{C}}$.*

Proof The canonical holomorphic structure of $T\Sigma$ is determined by a complex-linear Cauchy–Riemann type operator $\mathbf{D}_\Sigma : \Gamma(T\Sigma) \to \Gamma(\overline{\mathrm{End}}_{\mathbb{C}}(T\Sigma))$ that is the linearization at $\mathrm{Id}: (\Sigma, j) \to (\Sigma, j)$ of the nonlinear operator $\bar\partial_j \varphi := d\varphi + j \circ d\varphi \circ j \in \Gamma(\overline{\mathrm{Hom}}_{\mathbb{C}}(T\Sigma, \varphi^*T\Sigma))$ for maps $\varphi: \Sigma \to \Sigma$. It follows that a smooth vector field $X \in \Gamma(T\Sigma)$ is holomorphic near some point $z \in \Sigma$ if and only if it can be written as

$$X = \partial_\sigma\varphi_\sigma|_{\sigma=0}$$

for a smooth family of maps $\{\varphi_\sigma: \Sigma \to \Sigma\}_{\sigma\in(-\epsilon,\epsilon)}$ that satisfy $\varphi_0 = \mathrm{Id}$ and $\partial_\sigma\varphi_\sigma|_{\sigma=0} = X$ and are holomorphic near z. In this case, the maps

$$u_\sigma := u \circ \varphi_\sigma : \Sigma \to M$$

130 *Local Positivity of Intersections*

also satisfy $\bar{\partial}_J u_\sigma = 0$ in a neighborhood of z, and the section $\eta := \partial_\sigma u_\sigma|_{\sigma=0}$ $\in \Gamma(u^*TM)$ is related to the vector field X by $\eta = du(X)$. This implies that $\mathbf{D}_u\eta$ also vanishes near z. Since jX is also a holomorphic vector field near z, the same argument implies that the section $du(jX) = J\eta \in \Gamma(u^*TM)$ satisfies $\mathbf{D}_u(J\eta) = 0$ near z. Both of these facts together prove that $\mathbf{D}_u^{\mathbb{C}}\eta = 0$ vanishes near z. In summary, we've shown that du maps any locally defined holomorphic vector field to a locally defined section of u^*TM that is holomorphic with respect to $\mathbf{D}_u^{\mathbb{C}}$, and this is equivalent to $du: T\Sigma \to u^*TM$ being a holomorphic bundle map. $\qquad \square$

Proof of Proposition B.26 Since u is not constant and Σ is connected, the holomorphicity of $du \in \Gamma(\mathrm{Hom}_{\mathbb{C}}(T\Sigma, u^*TM))$ implies that zeroes of du and therefore also critical points of u are isolated. The definition of $T_u \subset u^*TM$ at immersed points of u is obvious; thus we only need to check that a smooth extension of this subbundle over the zero-set of du exists. Given a critical point $z_0 \in \Sigma$, choose a holomorphic local coordinate for Σ and a holomorphic trivialization of $\mathrm{Hom}_{\mathbb{C}}(T\Sigma, u^*TM)$ near z_0, so that du is expressed in this neighborhood as a $\mathbb{C}^n$-valued holomorphic function of the form

$$(z - z_0)^k F(z)$$

for some $k \in \mathbb{N}$ and a $\mathbb{C}^n$-valued holomorphic function F with $F(z_0) \neq 0$. We can then define T_u at each point z near z_0 to be the complex line in $T_{u(z)}M$ corresponding to the complex span of $F(z)$ in the trivialization. This definition matches the previous definition at the immersed points $z \neq z_0$ and thus makes $T_u \subset u^*TM$ into a smooth line bundle on a neighborhood of z_0, with the integer $k > 0$ as the order of the zero of $du \in \Gamma(\mathrm{Hom}_{\mathbb{C}}(T\Sigma, T_u))$ at z_0. $\qquad \square$

B.2.2 A Lemma on Normal Push-Offs

The message of the following result is that whenever $u: (\Sigma, j) \to (M, J)$ and $v: (\Sigma', j') \to (M, J)$ are two J-holomorphic curves related to each other by

$$\exp_{u \circ \varphi} \eta = v$$

for some diffeomorphism $\varphi: \Sigma' \to \Sigma$ and section η of φ^*N_u, the section η is subject to the similarity principle. For technical reasons, we will need to allow φ and η in the statement to have only finitely many derivatives, which forces nonsmooth bundles with nonsmooth Cauchy–Riemann type operators into the picture.

B.2 The Representation Formula

For convenience, we shall denote elements of the bundle $N_u \to \Sigma$ as pairs (z, w), where $z \in \Sigma$ and w belongs to the fiber $(N_u)_z$ over z. The zero-section thus consists of all pairs of the form $(z, 0)$, and there are canonical isomorphisms

$$T_{(z,0)}N_u = T_z\Sigma \oplus (N_u)_z \tag{B.13}$$

due to the natural identification of Σ with the zero-section and of vertical tangent spaces with fibers of N_u. Given a map $\varphi \colon \Sigma' \to \Sigma$, a section η of the induced bundle $\varphi^*N_u \to \Sigma'$ can now be written in the form $\eta(z) = (\varphi(z), f(z)) \in N_u$ with $f(z) \in (N_u)_{\varphi(z)}$ for $z \in \Sigma'$.

Proposition B.28 *Suppose $u \colon (\Sigma, j) \to (M, J)$ and $v \colon (\Sigma', j') \to (M, J)$ are smooth J-holomorphic curves, $\varphi \colon \Sigma' \to \Sigma$ is a diffeomorphism of class C^k for some $k \in \mathbb{N} \cup \{\infty\}$, $N_u \subset u^*TM$ is the generalized normal bundle of u in the sense of §B.2.1, $O \subset N_u$ is an open neighborhood of the zero-section, and*

$$\Psi \colon O \to M$$

is a smooth map that satisfies

$$\Psi(z, 0) = u(z) \quad \text{and} \quad d\Psi(z, 0)X = X \quad \text{for all } z \in \Sigma \text{ and } X \in (N_u)_z,$$

where the second condition makes sense due to the canonical splitting (B.13) and the inclusion $(N_u)_z \subset T_{u(z)}M$. If $\eta \colon \Sigma' \to \varphi^*N_u$ is a section of class C^k of the bundle $\varphi^*N_u \to \Sigma'$ with image in φ^*O such that

$$v(z) = \Psi(\varphi(z), \eta(z)) \quad \text{for all } z \in \Sigma',$$

then η satisfies $\mathbf{D}\eta = 0$ for some real-linear Cauchy–Riemann type operator $\mathbf{D}$ of class C^{k-1} on the bundle $(\varphi^*N_u, J) \to (\Sigma', j')$.

Proof Choose connections on the bundles TM and N_u. The induced bundle φ^*N_u is of class C^k, and the connection on N_u induces a connection on φ^*N_u of class C^{k-1} (cf. Remark B.5), whose covariant derivative operator we will denote by ∇. For $(z, w) \in O$, let

$$P_{(z,w)} \colon T_{u(z)}M \to T_{\Psi(z,w)}M$$

denote the isomorphism defined via parallel transport along the path $[0, 1] \to M \colon \tau \mapsto \Psi(z, \tau w)$. The connection on N_u also determines natural isomorphisms

$$T_{(z,w)}N_u = T_z\Sigma \oplus (N_u)_z \tag{B.14}$$

for each $(z, w) \in N_u$, where the two factors correspond to the horizontal and vertical subspaces, respectively. We can then associate to each $(z, w) \in O$ the linear map

Local Positivity of Intersections

$$F(z, w) := P_{(z,w)}^{-1} \circ d\Psi(z, w) \in \text{Hom}_{\mathbb{R}}\left(T_z\Sigma \oplus (N_u)_z, T_{u(z)}M\right),$$

which depends smoothly on $(z, w) \in O$ and satisfies

$$F(z, 0) = du(z) \oplus \mathbb{1}.$$

If we fix $z \in \Sigma$, then $F(z, \cdot)$ is a smooth map from $O_z := O \cap (N_u)_z$ to a fixed vector space of linear maps, and thus satisfies

$$F(z, w) = F(z, 0) + \int_0^1 \frac{d}{d\tau}F(z, \tau w)\,d\tau = F(z, 0) + \left(\int_0^1 d_2 F(z, \tau w)\,d\tau\right)w$$

$$= (du(z) \oplus \mathbb{1}) + \widetilde{F}(z, w)w,$$

where the integral at the end of the first line is used to define a smooth family of linear maps

$$\widetilde{F}(z, w) : (N_u)_z \to \text{Hom}_{\mathbb{R}}\left(T_z\Sigma \oplus (N_u)_z, T_{u(z)}M\right)$$

parametrized by $(z, w) \in O$.

Similarly, we associate to each $(z, w) \in O$ another linear map

$$G(z, w) := P_{(z,w)}^{-1} \circ J(\Psi(z, w)) \circ P_{(z,w)} \in \text{End}_{\mathbb{R}}(T_{u(z)}M),$$

which again depends smoothly on $(z, w) \in O$ and has image in a fixed vector space if z is fixed. We then have

$$G(z, w) = G(z, 0) + \int_0^1 \frac{d}{d\tau}G(z, \tau w)\,d\tau = G(z, 0) + \left(\int_0^1 d_2 G(z, \tau w)\,d\tau\right)w$$

$$= J(u(z)) + \widetilde{G}(z, w)w,$$

where the integral in the first line defines

$$\widetilde{G}(z, w) : (N_u)_z \to \text{End}_{\mathbb{R}}(T_{u(z)}M),$$

another family of linear maps with smooth dependence on the parameter $(z, w) \in O$.

Now suppose $v : (\Sigma', j') \to (M, J)$ is J-holomorphic and $v(z) = \Psi(\varphi(z), \eta(z))$, where $\eta : \Sigma' \to \varphi^*N_u$ is a C^k-smooth section with image in φ^*O. Using the splitting (B.14) determined by the connection on N_u, we have for each $z \in \Sigma'$,

$$dv(z) = d\Psi(\varphi(z), \eta(z)) \circ (d\varphi(z), \nabla\eta(z))$$

$$= P_{(\varphi(z),\eta(z))} \circ F(\varphi(z), \eta(z)) \circ (d\varphi(z), \nabla\eta(z)).$$

The parallel transport isomorphisms $P_{(\varphi(z),\eta(z))} : T_{u(\varphi(z))}M \to T_{v(z)}M$ now define a C^k-smooth real-linear bundle isomorphism $(u \circ \varphi)^*TM \to v^*TM$, and applying

B.2 The Representation Formula

its inverse to the nonlinear Cauchy–Riemann equation $dv + J(v) \circ dv \circ j' = 0$
gives an equation for real-linear bundle maps $T\Sigma \to \varphi^* u^* TM$,

$$
\begin{aligned}
0 &= F(\varphi, \eta) \circ (d\varphi, \nabla\eta) + G(\varphi, \eta) \circ F(\varphi, \eta) \circ (d\varphi \circ j', \nabla\eta \circ j') \\
&= \left[(du(\varphi) \oplus \mathbb{1}) + \widetilde{F}(\varphi, \eta)\eta \right] \circ (d\varphi, \nabla\eta) \\
&\quad + \left[J(u(\varphi)) + \widetilde{G}(\varphi, \eta)\eta \right] \circ \left[(du(\varphi) \oplus \mathbb{1}) + \widetilde{F}(\varphi, \eta)\eta \right] \circ (d\varphi \circ j', \nabla\eta \circ j') \\
&= \left[d(u \circ \varphi) + J(u \circ \varphi) \circ d(u \circ \varphi) \circ j' \right] + \left[\nabla\eta + J(u \circ \varphi) \circ \nabla\eta \circ j' \right] \\
&\quad + \left[\widetilde{F}(\varphi, \eta)\eta \right] \circ (d\varphi, \nabla\eta) \\
&\quad + \left[\widetilde{G}(\varphi, \eta)\eta \right] \circ \left[d(u \circ \varphi) \circ j' + \nabla\eta \circ j' + \left(\widetilde{F}(\varphi, \eta)\eta \right) \circ (d\varphi \circ j', \nabla\eta \circ j') \right] \\
&\quad + J(u \circ \varphi) \circ \left[\widetilde{F}(\varphi, \eta)\eta \right] \circ (d\varphi \circ j', \nabla\eta \circ j').
\end{aligned}
$$

Since η and φ are of class C^k, all terms in this expression are at least C^{k-1}-smooth functions of z, and each term in the last three lines can be understood as a product of C^{k-1}-smooth sections of various bundles, at least one of which is always of the form $B\eta$ for a C^{k-1}-smooth linear bundle map B from $\varphi^* N_u$ to some other bundle; e.g., we have $B(z) = \widetilde{F}(\varphi(z), \eta(z))$ in the first of these three lines and $B(z) = \widetilde{G}(\varphi(z), \eta(z))$ in the second. We can therefore abbreviate the last three lines as $\widehat{A}\eta$ for some C^{k-1}-smooth bundle map $\widehat{A} \colon \varphi^* N_u \to \mathrm{Hom}_{\mathbb{R}}(T\Sigma, \varphi^* u^* TM)$, so that the entire equation becomes

$$
\left[d(u \circ \varphi) + J(u \circ \varphi) \circ d(u \circ \varphi) \circ j' \right] + \left[\nabla\eta + J(u \circ \varphi) \circ \nabla\eta \circ j' \right] + \widehat{A}\eta = 0.
$$

Finally, let $\pi_N \colon u^* TM \to N_u$ denote the smooth bundle map defined by projecting $u^* TM = T_u \oplus N_u$ along T_u, which induces a C^k-smooth bundle map

$$
\pi_N \colon \varphi^* u^* TM \to \varphi^* N_u.
$$

In terms of the splitting $\varphi^* u^* TM = \varphi^* T_u \oplus \varphi^* N_u$, the term $d(u \circ \varphi) + J(u \circ \varphi) \circ d(u \circ \varphi) \circ j'$ in the above expression has image in $\varphi^* T_u$, while $\nabla\eta + J(u \circ \varphi) \circ \nabla\eta \circ j'$ has image in $\varphi^* N_u$. Applying π_N to the whole equation thus gives rise to

$$
\nabla\eta + J(u) \circ \nabla\eta \circ j' + \pi_N \widehat{A}\eta = 0.
$$

This is not quite yet a Cauchy–Riemann type equation; for this we would need the target of the bundle map $\pi_N \widehat{A}$ to be the bundle of complex-antilinear maps $\overline{\mathrm{Hom}}_{\mathbb{C}}(T\Sigma, \varphi^* N_u)$, whereas $\pi_N \widehat{A}$ sends N_u to the larger bundle $\mathrm{Hom}_{\mathbb{R}}(T\Sigma, \varphi^* N_u)$. We can fix this simply by taking the complex-antilinear part; i.e., we define

134   *Local Positivity of Intersections*

$$\varphi^* N_u \overset{A}{\to} \overline{\mathrm{Hom}}_{\mathbb{C}}((T\Sigma, j'), (\varphi^* N_u, J)),$$

$$Aw := \frac{1}{2}\left(\pi_N \widehat{A}w + J \circ \pi_N \widehat{A}w \circ j'\right).$$

Since $\pi_N \widehat{A}\eta = -\nabla\eta - J(u \circ \varphi) \circ \nabla\eta \circ j'$ and the latter is manifestly complex anti-linear, we have $A\eta = \pi_N \widehat{A}\eta$, proving that η also satisfies the Cauchy–Riemann type equation

$$\nabla\eta + J(u \circ \varphi) \circ \nabla\eta \circ j' + A\eta = 0. \qquad \square$$

B.2.3 Local Coordinates

For the rest of this section, we focus explicitly on the situation described in the statement of Theorem B.23. Our first objective is to find suitable coordinate charts near $z_0 \in \Sigma$ and $u(z_0) \in M$ so that u near z_0 becomes a map of the form

$$\mathbb{D}_\rho \to \mathbb{C} \times \mathbb{C}^n : z \mapsto (z^k, \widehat{u}(z))$$

for some $k \in \mathbb{N}$ with $\widehat{u}(z) = O(|z|^{k+1})$. In light of our discussion of the generalized tangent bundle $T_u \subset u^*TM$ in §B.2.1, it should be clear that the complex subspace $L \subset T_{u(z_0)}M$ mentioned in the theorem will be

$$L = (T_u)_{z_0}.$$

For the smooth coordinates near $u(z_0)$ on M, we impose the following conditions, which depend on the point $u(z_0) \in M$ and the subspace $(T_u)_{z_0}$, but not otherwise on the map $\Sigma \overset{u}{\to} M$:

(1) The point $u(z_0) \in M$ is identified with $0 \in \mathbb{C}^n$.
(2) The complex subspace $L \subset T_{u(z_0)}M$ is identified with $\mathbb{C} \times \{0\} \subset \mathbb{C}^n$.
(3) The map $u_0(z) := (z, 0) \in \mathbb{C} \times \mathbb{C}^{n-1}$ is J-holomorphic on $\mathbb{D}_\rho$ for sufficiently small $\rho > 0$, and J along the image of this map is identified with the standard complex structure i on $\mathbb{C}^n$.

Note that while the first two conditions are easy to achieve, the third is highly nontrivial. It is possible due to the standard local existence result for J-holomorphic curves with a fixed tangent vector – the latter follows from the implicit function theorem after performing a local rescaling argument to view $\bar{\partial}_J$ as a small perturbation of the surjective linear operator $\bar{\partial}$ (see, e.g., [Wena, Chapter 2] or [Sik94, Theorem 3.1.1]). After choosing a suitable J-holomorphic disk $\mathbb{D}_\rho \hookrightarrow M$ through $u(z_0)$, one can construct the desired coordinates by exponentiating in complex normal directions from this disk. With this understood, for the rest of this section we shall fix a choice of holomorphic coordinates near $z_0 \in \Sigma$ and smooth coordinates near $u(z_0) \in M$ as

B.2 The Representation Formula
135

described above in order to assume $(\Sigma, j) = (\mathbb{D}_\rho, i)$ with $z_0 = 0 \in \mathbb{D}_\rho$ for some $\rho > 0$, while J is a smooth almost complex structure on $\mathbb{C}^n = \mathbb{C} \times \mathbb{C}^{n-1}$ with $J(z, 0) = i$ for all $z \in \mathbb{D}_\rho$, and $u \colon (\mathbb{D}_\rho, i) \to (\mathbb{C}^n, J)$ is a J-holomorphic curve with $u(0) = 0$ and generalized tangent space $(T_u)_0 = \mathbb{C} \times \{0\}$.

We next seek a C^1-smooth coordinate change near the origin on the domain of u so that it becomes a map of the form $z \mapsto (z^k, O(|z|^{k+1}))$. We start with the observation that $u \colon (\mathbb{D}_\rho, i) \to (\mathbb{C}^n, J)$ itself satisfies the smooth complex-linear case of the similarity principle: indeed, the nonlinear Cauchy–Riemann equation

$$\partial_s u(z) + J(u(z)) \, \partial_t u(z) = 0$$

can be interpreted as a smooth complex-linear Cauchy–Riemann type equation $\mathbf{D}u = 0$ on the trivial rank n complex vector bundle over $\mathbb{D}_\rho$ with complex structure $\bar{J}(z) := J(u(z))$. As a consequence, Theorem B.20 gives

$$u(z) = \Phi(z)f(z)$$

on $\mathbb{D}_\rho$ after possibly shrinking $\rho > 0$, where $\Phi \colon \mathbb{D}_\rho \to \mathrm{GL}(2n, \mathbb{R})$ is the inverse of a smooth complex local trivialization and thus satisfies $\Phi(z) \circ i = J(u(z)) \circ \Phi(z)$, while $f \colon \mathbb{D}_\rho \to \mathbb{C}^n$ is a holomorphic function with $f(0) = 0$. Since $J(0) = i$, we can assume without loss of generality that $\Phi(0) = \mathbb{1}$. The assumption that u is not constant near z_0 implies in turn that f is nontrivial and thus satisfies

$$f(z) = z^k g(z)$$

for some $k \in \mathbb{N}$ and a holomorphic function $g \colon \mathbb{D}_\rho \to \mathbb{C}^n$ with $g(0) \neq 0$. By Corollary B.21 and Remark B.22, we can identify k as the degree of the lowest-order nontrivial term in the Taylor series of u at $z = 0$; equivalently, $k - 1$ is the vanishing order of $du \in \Gamma(\mathrm{Hom}_{\mathbb{C}}(T\Sigma, T_u))$ at $z = 0$, also known as the critical order of u at this point. The assumption $L = \mathbb{C} \times \{0\}$ now implies that after a complex-linear coordinate change on the domain, we may assume $g(0) = (1, 0) \in \mathbb{C} \times \mathbb{C}^{n-1}$. Thus $f(z) = (z^k g_1(z), z^{k+1} g_2(z))$ on $\mathbb{D}_\rho$ for some holomorphic functions $g_1 \colon \mathbb{D}_\rho \to \mathbb{C}$ and $g_2 \colon \mathbb{D}_\rho \to \mathbb{C}^{n-1}$, with $g_1(0) = 1$. Let us use the splitting $\mathbb{C}^n = \mathbb{C} \times \mathbb{C}^{n-1}$ to write $\Phi(z)$ in block form as

$$\Phi(z) = \begin{pmatrix} \mathbb{1} + \alpha(z) & \beta(z) \\ \gamma(z) & \mathbb{1} + \delta(z) \end{pmatrix},$$

where the blocks $\alpha(z)$, $\beta(z)$, $\gamma(z)$ and $\delta(z)$ are all regarded as *real*-linear maps between complex vector spaces, and all of them vanish at $z = 0$ since $\Phi(0) = \mathbb{1}$. Now $u(z)$ takes the form $(u_1(z), u_2(z)) \in \mathbb{C} \times \mathbb{C}^{n-1}$, where

Local Positivity of Intersections

$$u_1(z) = z^k g_1(z) + \alpha(z)z^k g_1(z) + \beta(z)z^{k+1} g_2(z),$$
$$u_2(z) = \gamma(z)z^k g_1(z) + (1 + \delta(z))z^{k+1} g_2(z). \tag{B.15}$$

We claim that, after shrinking $\rho > 0$ further if necessary, there exists a C^1-smooth function $\xi \colon \mathbb{D}_\rho \to \mathbb{C}$ such that $\xi(0) = 0$, $d\xi(0) = 1$ and $[\xi(z)]^k = u_1(z)$. Indeed, the desired function can be written for $z \neq 0$ as

$$\xi(z) = z \sqrt[k]{g_1(z) + \frac{1}{z^k}\alpha(z)z^k g_1(z) + \frac{1}{z^k}\beta(z)z^{k+1} g_2(z)},$$

where the expression under the root lies in a neighborhood of $g_1(0) = 1$ for z near 0; hence the root is uniquely defined as a continuous function of z on $\mathbb{D}_\rho$ if we set $\sqrt[k]{1} := 1$. It is clear that ξ is also smooth for $z \neq 0$ and differentiable at $z = 0$ with $d\xi(0) = 1$. Moreover, the fact that α and β are smooth functions vanishing at $z = 0$ implies that both are $O(|z|)$, so that the first derivative of the expression under the root is bounded on $\mathbb{D}_\rho \backslash \{0\}$. This is enough information to prove that $d\xi$ is also continuous at $z = 0$, so ξ is of class C^1.

Denote the inverse of the local C^1-diffeomorphism $z \mapsto \xi(z)$ by

$$\varphi := \xi^{-1} \colon \mathbb{D}_\rho \to \mathbb{D},$$

where we can again shrink $\rho > 0$ if necessary to make sure that φ is well defined and has image contained in the domain of u. The composition $u \circ \varphi$ is then well defined and satisfies

$$u \circ \varphi(z) = (z^k, \widehat{u}(z)),$$

where $\widehat{u} := u_2 \circ \varphi \colon \mathbb{D}_\rho \to \mathbb{C}^{n-1}$ is a C^1-smooth function satisfying the relation $\widehat{u}(\xi(z)) = u_2(z)$.

Lemma B.29 *The function $\widehat{u} \in C^1(\mathbb{D}_\rho, \mathbb{C}^{n-1})$ satisfies $\widehat{u}(z) = O(|z|^{k+1})$ and $d\widehat{u}(z) = O(|z|^k)$.*

Proof We have $u_2(z) = O(|z|^{k+1})$ by (B.15) since $\gamma(z)$ is a smooth function with $\gamma(0) = 0$, so, in particular, the first nontrivial term in the Taylor series of u_2 about $z = 0$ has degree at least $k + 1$, implying a similar conclusion for du_2 and thus $du_2(z) = O(|z|^k)$. The conditions $\varphi(0) = 0$ and $d\varphi(0) = 1$ imply also that $\varphi(z) = z + o(|z|)$. Writing $u_2(z) = |z|^{k+1} B(z)$ for a bounded function $B(z)$ near $z = 0$ and $\varphi(z) = z + |z| \cdot r(z)$ for a remainder function with $\lim_{z \to 0} r(z) = 0$, we find

$$\widehat{u}(z) = u_2(\varphi(z)) = u_2(z + |z| \cdot r(z)) = \left| z + |z| \cdot r(z) \right|^{k+1} B(z + |z| \cdot r(z))$$

$$= |z|^{k+1} \cdot \left| \frac{z}{|z|} + r(z) \right|^{k+1} B(z + |z| \cdot r(z)) = O(|z|^{k+1}).$$

B.2 The Representation Formula

Similarly, $d\widehat{u}(z) = du_2(\varphi(z)) \circ d\varphi(z)$, where $d\varphi$ is continuous and therefore bounded near $z = 0$, and the same argument as above gives $du_2(\varphi(z)) = O(|z|^k)$ since $du_2(z) = O(|z|^k)$, so the result for $d\widehat{u}(z)$ follows. $\qquad\qquad\square$

Now if $v\colon (\Sigma', j') \to (\mathbb{C}^n, J)$ is a second J-holomorphic curve with a point $\zeta_0 \in \Sigma'$ such that $v(\zeta_0) = u(z_0) = 0$, $(T_v)_{\zeta_0} = (T_u)_{z_0} = \mathbb{C} \times \{0\}$ and the critical orders at $v(\zeta_0)$ and $u(z_0)$ match, then we can repeat the same argument to find a C^1-smooth local diffeomorphism ψ from $\mathbb{D}_\rho$ to a neighborhood of ζ_0 in Σ' sending $0 \mapsto \zeta_0$ such that

$$v \circ \psi(z) = (z^k, \widehat{v}(z)),$$

with

$$\widehat{v} \in C^1(\mathbb{D}_\rho, \mathbb{C}^{n-1}) \quad \text{such that} \quad \widehat{v}(z) = O(|z|^{k+1}) \text{ and } d\widehat{v}(z) = O(|z|^k).$$

The main goal for the rest of this section is to prove that the C^1-smooth function

$$h(z) = (0, \widehat{h}(z)) := (0, \widehat{v}(z) - \widehat{u}(z)) = v \circ \psi(z) - u \circ \varphi(z)$$

is either identically zero or satisfies the formula $h(z) = z^\ell C + o(|z|^\ell)$ for some $C \in \mathbb{C}^n \setminus \{0\}$ and $\ell > k$.

Remark B.30 It should be emphasized that φ and ψ are, in general, neither holomorphic nor smooth, so $u \circ \varphi$ and $v \circ \psi$ are pseudoholomorphic curves of class C^1 with respect to complex structures on $\mathbb{D}_\rho$ that are nonstandard and continuous but not generally smooth, though since $d\varphi(0) = d\psi(0) = \mathbb{1}$ and φ and ψ are smooth outside the origin, both complex structures are standard at the origin and smooth elsewhere. As a special case, however, we could take v to be

$$v\colon (\mathbb{D}_\rho, i) \to (\mathbb{C} \times \mathbb{C}^{n-1}, J)\colon z \mapsto (z^k, 0),$$

which is J-holomorphic due to the third condition imposed on our local coordinates in M. The claim in Theorem B.23 that $\widehat{u}$ and $\widehat{v}$ each satisfy formulas of the form $z^\ell C + o(|z|^\ell)$ will thus follow as a special case of the general formula for $\widehat{v} - \widehat{u}$.

B.2.4 Constructing the Normal Push-Off

We now define a neighborhood $O \subset N_u$ and a map $\Psi\colon O \to M = \mathbb{C}^n$ as in Proposition B.28. Our first task is to specify a concrete complex subbundle $N_u \subset u^*TM$ complementary to T_u. Since $(T_u)_0 = \mathbb{C} \times \{0\}$, any complex subbundle that matches $\{0\} \times \mathbb{C}^{n-1}$ at 0 will do if we are willing to shrink $\rho > 0$, as the two subbundles will necessarily be transverse on $\mathbb{D}_\rho$ for ρ sufficiently small. Let

138   *Local Positivity of Intersections*

$$e_1, \ldots, e_n \in \mathbb{C}^n$$

denote the standard complex basis of $\mathbb{C}^n$, and for each $w = (x_2 + iy_2, \ldots, x_n + iy_n) \in \mathbb{C}^{n-1}$, define a smooth vector field on $\mathbb{C}^n$ by

$$X_w(p) := \sum_{j=2}^{n} \left(x_j e_j + y_j J(p) e_j \right).$$

Since $J(z, 0) = i$ for $(z, 0) \in \mathbb{D}_\rho \times \mathbb{C}^{n-1}$, at such points we have $X_w(z, 0) = (0, w) \in \mathbb{C}^n$ for all $w \in \mathbb{C}^n$. We shall regard $N_u \to \mathbb{D}_\rho$ in the following as the pullback along $u \colon \mathbb{D}_\rho \to \mathbb{C}^n$ of the smooth subbundle of $T\mathbb{C}^n$ spanned by the vector fields X_w for all $w \in \mathbb{C}^{n-1}$. This bundle comes equipped with a global trivialization

$$N_u \to \mathbb{D}_\rho \times \mathbb{C}^{n-1} \colon X_w(u(z)) \mapsto (z, w). \tag{B.16}$$

For a constant $\delta > 0$, we define the open set

$$O_\delta := \left\{ (z, w) \mid |w| < \delta \right\} \subset \mathbb{D}_\rho \times \mathbb{C}^{n-1}$$

and smooth map

$$\Psi \colon O_\delta \to \mathbb{C}^n \colon (z, w) \mapsto u(z) + X_w(u(z)).$$

In light of the trivialization (B.16), we can equivalently regard O_δ as a neighborhood of the zero-section in N_u on which Ψ is defined as in Proposition B.28. Notice that $\Psi(\varphi(z), 0) = u \circ \varphi(z) = (z^k, \hat{u}(z))$. The goal is to apply Proposition B.28 to the following construction:

Lemma B.31 *Choosing $\delta > 0$ sufficiently small and then shrinking $\rho > 0$ further if necessary, there exist C^1-smooth functions $\theta \colon \mathbb{D}_\rho \to \mathbb{C}$ and $\eta \colon \mathbb{D}_\rho \to \mathbb{C}^{n-1}$ such that $\theta(0) = 0$, $d\theta(0) = \mathbb{1}$, $\eta(z) = O(|z|^{k+1})$, and*

$$v \circ \psi(z) = \Psi(\varphi \circ \theta(z), \eta(z)) \quad \text{for all} \quad z \in \mathbb{D}_\rho.$$

The proof of this lemma requires some preparation. We will use the notation d_1 and d_2 to denote the differentials of Ψ or X_w with respect to the first variable $z \in \mathbb{C}$ or second variable $w \in \mathbb{C}^{n-1}$, respectively, e.g., writing

$$d_1\Psi(z, w) \in \mathrm{Hom}_{\mathbb{R}}(\mathbb{C}, \mathbb{C}^n), \qquad d_2\Psi(z, w) \in \mathrm{Hom}_{\mathbb{R}}(\mathbb{C}^{n-1}, \mathbb{C}^n).$$

Let us also write

$$\Psi(z, w) =: (\check{\Psi}(z, w), \widehat{\Psi}(z, w)) \in \mathbb{C} \times \mathbb{C}^{n-1} \quad \text{and}$$

$$X_w(p) = (\check{X}_w(p), \widehat{X}_w(p)) \in \mathbb{C} \times \mathbb{C}^{n-1},$$

so $\check{\Psi}(z, w) = u_1(z) + \check{X}_w(u(z))$ and $\widehat{\Psi}(z, w) = u_2(z) + \widehat{X}_w(u(z))$.

B.2 The Representation Formula 139

Lemma B.32 *Given a compact region $K \subset \mathbb{C} \times \mathbb{C}^{n-1}$, there exists a constant $C > 0$ such that the following estimates hold for all $(z, p) \in K$ and all $w \in \mathbb{C}^{n-1}$:*

$$\left| \check{X}_w(z, p) \right| \leq C|p| \cdot |w|, \quad \left| \hat{X}_w(z, p) - w \right| \leq C|p| \cdot |w|, \quad |d_1 X_w(z, p)| \leq C|p| \cdot |w|.$$

Proof For each $(z, p) \in \mathbb{C} \times \mathbb{C}^{n-1}$, $w \mapsto \check{X}_w(z, p)$ defines a real-linear map $\check{X}(z, p) \colon \mathbb{C}^{n-1} \to \mathbb{C}$. Since $J(z, 0) = i$ for all z, we have $X_w(z, 0) = (0, w)$; thus $\check{X}(z, 0) = 0$, and the smoothness of $\check{X}_w(z, p)$ with respect to z and p then gives rise to an estimate

$$|\check{X}(z, p)| \leq C|p|$$

from which the first estimate above follows. The second estimate follows in the same manner since $\hat{X}(z, 0)$ is the identity map $\mathbb{1} \colon \mathbb{C}^{n-1} \to \mathbb{C}^{n-1}$; hence $|\hat{X}(z, p) - \hat{X}(z, 0)| \leq C|p|$. For the third estimate, one observes that $w \mapsto d_1 X_w(z, p)$ is also a real-linear map $\mathbb{C}^{n-1} \to \mathrm{Hom}_{\mathbb{R}}(\mathbb{C}, \mathbb{C}^n)$ for every (z, p) and $d_1 X_w(z, 0) = 0$ since $X_w(z, 0)$ is independent of z, so the same argument applies. $\qquad\square$

Due to the coordinate choices made in §B.2.3, we also have $u(z) = (u_1(z), u_2(z)) = (z^k, 0) + O(|z|^{k+1})$; thus $|u_1(z)| \geq c|z|^k$ and $|u_2(z)| \leq C|z|^{k+1}$ for some constants $c, C > 0$, where we are free to assume C is the same constant as in Lemma B.32. It follows that for $z \neq 0$,

$$\left| \check{\Psi}(z, w) \right| = \left| u_1(z) + \check{X}_w(u(z)) \right| \geq |u_1(z)| - \left| \check{X}_w(u_1(z), u_2(z)) \right| \geq c|z|^k$$
$$- C|u_2(z)| \cdot |w| \geq c|z|^k - C^2|z|^{k+1}|w| = |z|^k \left(c - C^2|w| \cdot |z| \right),$$

which is positive if $|w| < c/\rho C^2$. This proves the following lemma:

Lemma B.33 *If $\delta > 0$ is sufficiently small, then Ψ preserves the subset $\{(z, w) \in \mathbb{C} \times \mathbb{C}^{n-1} \mid z \neq 0\}$.* $\qquad\square$

Now consider the C^1-smooth function $\Psi_1 = (\check{\Psi}_1, \hat{\Psi}_1) \colon O_\delta \to \mathbb{C} \times \mathbb{C}^n$ defined by

$$\Psi_1(z, w) = \Psi(\varphi(z), w) = u(\varphi(z)) + X_w(u(\varphi(z)))$$
$$= \left(z^k + \check{X}_w(z^k, \hat{u}(z)), \hat{u}(z) + \hat{X}_w(z^k, \hat{u}(z)) \right),$$

and extend this to a C^1-smooth family of maps $\Psi_\varepsilon = (\check{\Psi}_\varepsilon, \hat{\Psi}_\varepsilon) \colon O_\delta \to \mathbb{C} \times \mathbb{C}^{n-1}$ for $0 < \varepsilon \leq 1$ by

$$\Psi_\varepsilon(z, w) = \left(\frac{\breve{\Psi}_1(\varepsilon z, w)}{\varepsilon^k}, \widehat{\Psi}_1(\varepsilon z, w) \right)$$

$$= \left(z^k + \frac{\breve{X}_w(\varepsilon^k z^k, \widehat{u}(\varepsilon z))}{\varepsilon^k}, \widehat{u}(\varepsilon z) + \widehat{X}_w(\varepsilon^k z^k, \widehat{u}(\varepsilon z)) \right).$$

We would like to understand what happens to Ψ_ε as $\varepsilon \to 0$, but from a slightly different vantage point, namely after transforming the first complex variable in $\mathbb{C} \times \mathbb{C}^{n-1}$ to holomorphic cylindrical coordinates. Define the biholomorphic map

$$f \colon \mathbb{R} \times S^1 \xrightarrow{\cong} \mathbb{C}\backslash\{0\} \colon (s, t) \mapsto e^{2\pi(s+it)},$$

let

$$\dot{O}_\delta := \left\{ (s, t, w) \in \mathbb{R} \times S^1 \times \mathbb{C}^{n-1} \,\middle|\, (f(s, t), w) \in O_\delta \right\},$$

and consider the family of C^1-smooth maps

$$\Psi'_\varepsilon := (\breve{\Psi}'_\varepsilon, \widehat{\Psi}'_\varepsilon) := (f^{-1} \times \mathrm{Id}) \circ \Psi_\varepsilon \circ (f \times \mathrm{Id}) \colon \dot{O}_\delta \to \mathbb{R} \times S^1 \times \mathbb{C}^{n-1},$$

which are given by

$$\breve{\Psi}'_\varepsilon(s, t, w) = f^{-1} \circ \breve{\Psi}_\varepsilon(e^{2\pi(s+it)}, w) \quad \text{and} \quad \widehat{\Psi}'_\varepsilon(s, t, w) = \widehat{\Psi}_\varepsilon(e^{2\pi(s+it)}, w).$$

Since $\widehat{\Psi}_\varepsilon(e^{2\pi(s+it)}, w) = \widehat{\Psi}_1(\varepsilon e^{2\pi(s+it)}, w)$, the functions $\widehat{\Psi}'_\varepsilon \colon \dot{O}_\delta \to \mathbb{C}^{n-1}$ converge in C^1 to $\widehat{\Psi}'_0(s, t, w) := \widehat{\Psi}_1(0, w) = w$ as $\varepsilon \to 0$. The convergence of $\breve{\Psi}'_\varepsilon \colon \dot{O}_\delta \to \mathbb{R} \times S^1$ as $\varepsilon \to 0$ will be deduced from the next lemma. To motivate the hypotheses in this statement, notice that the required estimates are satisfied automatically by any *smooth* function $g(z, w)$ that is of the form az^k plus terms that are higher order in z; the formulation below is only more complicated than this because we need to allow functions that are of class C^1 and not smooth.

Lemma B.34 *Fix $r \in \mathbb{R}$, $\rho := e^{2\pi r} > 0$ and an open set $\mathcal{U} \subset \mathbb{R}^n$, and suppose*

$$g \colon \mathbb{D}_\rho \times \mathcal{U} \to \mathbb{C}$$

is a function of class C^1 satisfying $g(z, w) \neq 0$ for all $z \neq 0$, along with estimates of the form

$$|g(z, w) - az^k| \le C|z|^{k+1}, \qquad\qquad |d_2 g(z, w)| \le C|z|^{k+1},$$

$$\left| \frac{\partial g}{\partial z}(z, w) - kaz^{k-1} \right| \le C|z|^k, \qquad\qquad \left| \frac{\partial g}{\partial \bar{z}}(z) \right| \le C|z|^k$$

for a constant $C > 0$ independent of $(z, w) \in \mathbb{D}_\rho \times \mathcal{U}$, where $a \in \mathbb{C}\backslash\{0\}$ and $k \in \mathbb{N}$ are constants, and $d_2 g(z, w) \colon \mathbb{R}^n \to \mathbb{C}$ denotes the differential with respect

B.2 The Representation Formula 141

to the second variable $w \in \mathcal{U}$. Using the biholomorphic map $f \colon \mathbb{R} \times S^1 \overset{\cong}{\to} \mathbb{C} \backslash \{0\} \colon (s, t) \mapsto e^{2\pi(s+it)}$, define for each $\varepsilon \in (0, 1]$ the maps $g_\varepsilon \colon \mathbb{D}_\rho \times \mathcal{U} \to \mathbb{C}$ and $g'_\varepsilon \colon (-\infty, r] \times S^1 \times \mathcal{U} \to \mathbb{R} \times S^1$ by

$$g_\varepsilon(z, w) := \frac{g(\varepsilon z, w)}{\varepsilon^k}, \quad \text{and} \quad g'_\varepsilon(s, t, w) := f^{-1} \circ g_\varepsilon(f(s, t), w).$$

Then as $\varepsilon \to 0$, the maps g'_ε are C^1-convergent on $(-\infty, r] \times S^1 \times \mathcal{U}$ to

$$g'_0(s, t, w) := (ks + s_0, kt + t_0),$$

where $e^{2\pi(s_0 + it_0)} = a$.

Proof By assumption, we can write

$$g(z, w) = az^k + |z|^{k+1} B(z, w), \qquad d_2 g(z, w) = |z|^{k+1} B_w(z, w),$$

$$\frac{\partial g}{\partial z}(z) = kaz^{k-1} + |z|^k B_z(z, w), \qquad \frac{\partial g}{\partial \bar{z}}(z) = |z|^k B_{\bar{z}}(z, w)$$

for bounded functions B, B_w, B_z and $B_{\bar{z}}$. Then

$$g_\varepsilon(z, w) = az^k + \varepsilon |z|^{k+1} B(\varepsilon z, w),$$

and on the punctured domain $\dot{\mathbb{D}}_\rho \times \mathcal{U}$ for $\dot{\mathbb{D}}_\rho := \mathbb{D}_\rho \backslash \{0\}$, we therefore have uniform convergence

$$\frac{g_\varepsilon(z, w)}{z^k} = a + \varepsilon |z| \frac{|z|^k}{z^k} B(\varepsilon z, w) \to a \quad \text{as} \quad \varepsilon \to 0.$$

We claim that this convergence is also in C^1 on $\dot{\mathbb{D}}_\rho \times \mathcal{U}$. Indeed, we have

$$d_2 \left(\frac{g_\varepsilon(z, w)}{z^k} \right) = \frac{1}{\varepsilon^k z^k} d_2 g(\varepsilon z, w) = \frac{1}{\varepsilon^k z^k} \varepsilon^{k+1} |z|^{k+1} B_w(\varepsilon z, w)$$

$$= \varepsilon |z| \frac{|z|^k}{z^k} B_w(\varepsilon z, w),$$

along with

$$\frac{\partial}{\partial z} \left(\frac{g_\varepsilon(z, w)}{z^k} \right) = \frac{1}{z^k} \frac{\partial}{\partial z} g_\varepsilon(z, w) - \frac{k}{z^{k+1}} g_\varepsilon(z, w)$$

$$= \frac{1}{\varepsilon^{k-1} z^k} \frac{\partial g}{\partial z}(\varepsilon z, w) - \frac{k}{\varepsilon^k z^{k+1}} g(\varepsilon z, w)$$

$$= \frac{1}{\varepsilon^{k-1} z^k} \left[\varepsilon^{k-1} kaz^{k-1} + \varepsilon^k |z|^k B_z(\varepsilon z, w) \right]$$

$$\quad - \frac{k}{\varepsilon^k z^{k+1}} \left[\varepsilon^k az^k + \varepsilon^{k+1} |z|^{k+1} B(\varepsilon z, w) \right]$$

$$= \varepsilon \left[\frac{|z|^k}{z^k} B_z(\varepsilon z, w) - k \frac{|z|^{k+1}}{z^{k+1}} B(\varepsilon z, w) \right],$$

142　　　*Local Positivity of Intersections*

and

$$\frac{\partial}{\partial \bar{z}}\left(\frac{g_\varepsilon(z,w)}{z^k}\right) = \frac{1}{z^k}\frac{\partial}{\partial \bar{z}}g_\varepsilon(z,w) = \frac{1}{\varepsilon^{k-1}z^k}\frac{\partial g}{\partial \bar{z}}(\varepsilon z,w) = \frac{1}{\varepsilon^{k-1}z^k}\varepsilon^k|z|^k B_{\bar{z}}(\varepsilon z,w)$$

$$= \varepsilon\frac{|z|^k}{z^k}B_{\bar{z}}(\varepsilon z,w).$$

All of these converge uniformly to 0 as $\varepsilon \to 0$.

To relate this to the maps g'_ε, identify $\mathbb{R} \times S^1$ with $\mathbb{C}/i\mathbb{Z}$ and write $f(\zeta) = e^{2\pi\zeta}$, so g'_ε is now determined by g_ε according to the formula $e^{2\pi g'_\varepsilon(\zeta,w)} = g_\varepsilon(z,w)$ for $z = e^{2\pi\zeta}$, implying

$$e^{2\pi[g'_\varepsilon(\zeta,w)-k\zeta]} = \frac{g_\varepsilon(z,w)}{z^k}.$$

For ε small enough, the convergence we have just established above implies that the right-hand side lies in a compact neighborhood of $a \in \mathbb{C}\backslash\{0\}$ on which the holomorphic logarithm function can be defined, giving rise to the formula

$$g'_\varepsilon(\zeta,w) - k\zeta = \frac{1}{2\pi}\log\left(g_\varepsilon(z,w)/z^k\right).$$

The right-hand side is C^1-convergent to the constant $\frac{1}{2\pi}\log(a)$ when regarded as a function of $(z,w) \in \dot{\mathbb{D}}_\rho \times \mathcal{U}$, and composing it with the transformation $(\zeta,w) \mapsto (e^{2\pi\zeta},w)$ in order to view it as a function of $(\zeta,w) \in (-\infty,r] \times S^1 \times \mathcal{U}$ does not change this result; thus we obtain C^1-convergence of g'_ε to $k\zeta + \frac{1}{2\pi}\log(a)$. □

We would now like to feed the function

$$\check{\Psi}_1(z,w) = z^k + \check{X}_w(z^k,\hat{u}(z))$$

into Lemma B.34. Since $\hat{u}(z) = O(|z|^{k+1})$, Lemma B.32 implies

$$\left|\check{\Psi}_1(z,w) - z^k\right| \le C|w| \cdot |\hat{u}(z)| \le C'|w| \cdot |z|^{k+1}$$

for some constant $C' > 0$ independent of z and w, and another application of Lemma B.32 together with the fact that $\check{\Psi}_1(z,w)$ depends linearly on w gives

$$\left|d_2\check{\Psi}_1(z,w)w'\right| = \left|\check{X}_{w'}(z^k,\hat{u}(z))\right| \le C|\hat{u}(z)| \cdot |w'|;$$

hence

$$\left|d_2\check{\Psi}_1(z,w)\right| \le C|\hat{u}(z)| \le C'|z|^{k+1}.$$

For the required estimates on derivatives with respect to z, it will suffice to prove that the function

$$\xi_w(z) := \check{X}_w(z^k,\hat{u}(z)) \quad \text{satisfies} \quad |d\xi_w(z)| \le C|z|^k$$

B.2 The Representation Formula 143

for a constant $C > 0$ independent of $(z, w) \in O_\delta$. We have $|d\widehat{u}(z)| = O(|z|^k)$ by Lemma B.29 and can assume $d_2 \check{X}_w(z^k, \widehat{u}(z))$ is bounded for $(z, w) \in O_\delta$, so applying the third estimate in Lemma B.32 gives

$$|d\xi_w(z)| = \left| d_1 \check{X}_w(z^k, \widehat{u}(z)) \circ (kz^{k-1}) + d_2 \check{X}_w(z^k, \widehat{u}(z)) \circ d\widehat{u}(z) \right|$$
$$\leq C|w| \cdot |\widehat{u}(z)| \cdot |z|^{k-1} + C|z|^k \leq C'|z|^{2k} + C|z|^k = O(|z|^k).$$

We can now apply Lemma B.34 and conclude the following:

Lemma B.35 *The maps* $\Psi'_\varepsilon : \dot{O}_\delta \to \mathbb{R} \times S^1 \times \mathbb{C}^{n-1}$ *are* C^1-*convergent as* $\varepsilon \to 0$ *to*

$$\Psi'_0(s, t, w) := (ks, kt, w). \qquad \square$$

A crucial detail in Lemma B.35 is that the C^1-convergence is not just on compact subsets, but remains uniform (including first derivatives) as s varies on the unbounded half-interval $(-\infty, r]$. We conclude from this, in particular that for all $\varepsilon > 0$ sufficiently small, Ψ'_ε is a local C^1-diffeomorphism whose image contains the set

$$\dot{O}'_\delta := \left\{ (s, t, w) \in \dot{O}_\delta \mid |w| < \delta/2 \right\}.$$

This is finally enough information to prove the main result of this subsection.

Proof of Lemma B.31 Denote $v_1(z) = (\check{v}(z), \widehat{v}(z)) := v \circ \psi(z)$, so $\check{v}(z) = z^k$. Our objective is to find a suitable local C^1-diffeomorphism $\theta : \mathbb{D}_\rho \to \mathbb{C}$ sending $0 \mapsto 0$ and a C^1-function $\eta : \mathbb{D}_\rho \to \mathbb{C}^{n-1}$ such that the relation

$$\Psi_1(\theta(z), \eta(z)) = v_1(z) \tag{B.17}$$

holds if the disk $\mathbb{D}_\rho$ is taken to be sufficiently small. We will do this by applying the same rescaling and cylindrical transformations to θ, η and v_1 that were applied above for Ψ_1, as the existence of such functions for $\varepsilon > 0$ sufficiently small will become obvious in cylindrical coordinates due to the convergence $\Psi'_\varepsilon \to \Psi'_0$.

Concretely, if maps θ and η as in (B.17) were already known, then, for $\varepsilon \in (0, 1]$, we could define $\theta_\varepsilon : \mathbb{D}_\rho \to \mathbb{C}, \eta_\varepsilon : \mathbb{D}_\rho \to \mathbb{C}^{n-1}$ and $v_\varepsilon : \mathbb{D}_\rho \to \mathbb{C} \times \mathbb{C}^{n-1}$ by

$$\theta_\varepsilon(z) := \frac{\theta(\varepsilon z)}{\varepsilon}, \quad \eta_\varepsilon(z) := \eta(\varepsilon z), \quad v_\varepsilon(z) := \left(\frac{\check{v}(\varepsilon z)}{\varepsilon^k}, \widehat{v}(\varepsilon z) \right) = (z^k, \widehat{v}(\varepsilon z)),$$

which must then satisfy the relation

$$\Psi_\varepsilon(\theta_\varepsilon(z), \eta_\varepsilon(z)) = v_\varepsilon(z). \tag{B.18}$$

Transforming one step further, let us again identify $\mathbb{R} \times S^1$ with $\mathbb{C}/i\mathbb{Z}$ and write $f(\zeta) = e^{2\pi\zeta}, \rho = 2\pi r$. If θ is a local diffeomorphism sending $0 \mapsto 0$, then

144 *Local Positivity of Intersections*

we can assume $\theta_\varepsilon(z) \neq 0$ for all $z \neq 0$ and $\varepsilon > 0$ sufficiently small, and can therefore define maps $\theta'_\varepsilon \colon (-\infty, r] \times S^1 \to \mathbb{R} \times S^1$, $\eta'_\varepsilon \colon (-\infty, r] \times S^1 \to \mathbb{C}^{n-1}$ and $v'_\varepsilon \colon (-\infty, r] \times S^1 \to \mathbb{R} \times S^1 \times \mathbb{C}^{n-1}$ by

$$\theta'_\varepsilon := f^{-1} \circ \theta_\varepsilon \circ f, \qquad \eta'_\varepsilon := \eta_\varepsilon \circ f, \qquad v'_\varepsilon := (f^{-1} \times \mathrm{Id}) \circ v_\varepsilon \circ f.$$

This last map is of the form

$$v'_\varepsilon(\zeta) = (k\zeta, \widehat{v}(\varepsilon e^{2\pi\zeta})),$$

and thus for $\varepsilon \to 0$ we have C^1-convergence $v'_\varepsilon \to v'_0$, where

$$v'_0(\zeta) := (k\zeta, 0) = \Psi'_0(\zeta, 0).$$

The cylindrical coordinate version of (B.18) is now the relation

$$\Psi'_\varepsilon(\theta'_\varepsilon(\zeta), \eta'_\varepsilon(\zeta)) = v'_\varepsilon(\zeta), \tag{B.19}$$

which is equivalent to (B.18) for each $\varepsilon > 0$.

The discussion of θ and η has been purely hypothetical thus far, but we are now in a position to find actual maps θ'_ε and η'_ε such that (B.19) is satisfied. Indeed, after shifting the upper boundary of the half-cylinder $(-\infty, r] \times S^1$ slightly if necessary, the convergence of local C^1-diffeomorphisms $\Psi'_\varepsilon \to \Psi'_0$ together with the convergence $v'_\varepsilon \to v'_0$ implies that, for every $\zeta \in (-\infty, r] \times S^1$, there exists a unique continuous family of points $(\theta'_\varepsilon(\zeta), \eta'_\varepsilon(\zeta)) \in \dot{O}'_\delta$ for $\varepsilon \geq 0$ sufficiently small such that (B.19) holds and $(\theta_0(\zeta), \eta_0(\zeta)) = (\zeta, 0)$; notice that the $\varepsilon = 0$ case of (B.19) is then the relation $v'_0(\zeta) = \Psi'_0(\zeta, 0)$ already established. Since the Ψ_ε are local C^1-diffeomorphisms for $\varepsilon \geq 0$ small and v_ε is of class C^1, the maps θ_ε and η_ε defined in this way are also of class C^1 and form a C^1-continuous family with respect to the parameter ε, implying, in particular, that we have C^1-convergence $\theta_\varepsilon \to \theta_0$ and $\eta_\varepsilon \to 0$ as $\varepsilon \to 0$. To obtain the actual objective, we only need fix $\varepsilon > 0$ sufficiently small and observe that both of the transformations $(\theta, \eta) \mapsto (\theta_\varepsilon, \eta_\varepsilon)$ and $(\theta_\varepsilon, \eta_\varepsilon) \mapsto (\theta'_\varepsilon, \eta'_\varepsilon)$ described above are reversible, at least if we are willing to restrict the domain of θ and η to a *punctured* disk $\dot{\mathbb{D}}_\rho$ whose size is reduced in proportion to the size of ε. After this reversal, we have a pair of C^1-smooth maps $\theta \colon \dot{\mathbb{D}}_\rho \to \dot{\mathbb{C}}$ and $\eta \colon \dot{\mathbb{D}}_\rho \to \mathbb{C}^{n-1}$ that satisfy (B.17) on the punctured disk $\dot{\mathbb{D}}_\rho$.

We claim that both θ and η can be extended over the puncture to functions of class C^1 on $\mathbb{D}_\rho$, with

$$\theta(0) = 0, \quad d\theta(0) = \mathbb{1}, \qquad \text{and} \qquad \eta(0) = 0, \quad d\eta(0) = 0.$$

For θ, we consider the functions $g_\varepsilon(z) := \frac{\theta_\varepsilon(z)}{z}$ on $\dot{\mathbb{D}}_\rho$ and observe that since θ'_ε converges in C^1 on $(-\infty, r] \times S^1$ to $\theta'_0(\zeta) = \zeta$,

$$g_\varepsilon \circ f(\zeta) = e^{2\pi[\theta'_\varepsilon(\zeta) - \zeta]}$$

B.2 The Representation Formula 145

is C^1-convergent on $(-\infty, r] \times S^1$ to the constant function with value 1. This implies that g_ε converges uniformly on $\dot{\mathbb{D}}_\rho$ to 1, and, writing $\theta_\varepsilon(z) = \theta(\varepsilon z)/\varepsilon$, we obtain the relation

$$\theta(\varepsilon z) = \varepsilon z g_\varepsilon(z) = \varepsilon z + \varepsilon z \left[g_\varepsilon(z) - 1\right]$$

for all $z \in \dot{\mathbb{D}}_\rho$. If we restrict this relation to points z on the boundary of $\mathbb{D}_\rho$ and introduce a new variable $w := \varepsilon z$ living in a neighborhood of $0 \in \mathbb{D}_\rho$, we can define a remainder function

$$R(w) := \frac{w}{|w|} \left[g_{|w|/\rho}(z) - 1\right]$$

that satisfies $\lim_{w \to 0} R(w) = 0$ due to the uniform convergence of g_ε, and it turns the foregoing relation into $\theta(w) = w + |w|R(w)$. Defining $\theta(0) := 0$ therefore makes θ continuous and differentiable at 0, with $d\theta(0) = \mathbb{1}$.

To prove that $d\theta(z)$ is also continuous at $z = 0$, we use the uniform convergence of the first derivatives of $g_\varepsilon \circ f$: writing $z = f(\zeta) = e^{2\pi\zeta}$, this convergence implies

$$\frac{\partial}{\partial \zeta} g_\varepsilon \circ f(\zeta) = \frac{\partial g_\varepsilon}{\partial z} \frac{\partial z}{\partial \zeta} = 2\pi z \frac{\partial g_\varepsilon}{\partial z} \to 0, \quad \text{and}$$

$$\frac{\partial}{\partial \bar{\zeta}} g_\varepsilon \circ f(\zeta) = \frac{\partial g_\varepsilon}{\partial \bar{z}} \frac{\partial \bar{z}}{\partial \bar{\zeta}} = 2\pi \bar{z} \frac{\partial g_\varepsilon}{\partial \bar{z}} \to 0$$

as $\varepsilon \to 0$. From the convergence of $\bar{z}\frac{\partial g_\varepsilon}{\partial \bar{z}}$, we obtain

$$\bar{z}\frac{\partial}{\partial \bar{z}} \left(\frac{\theta_\varepsilon(z)}{z}\right) = \frac{\bar{z}}{z} \frac{\partial}{\partial \bar{z}} \theta_\varepsilon(z) = \frac{\bar{z}}{z} \frac{\partial \theta}{\partial \bar{z}}(\varepsilon z) \to 0,$$

implying $\lim_{z \to 0} \frac{\partial \theta}{\partial \bar{z}}(z) = 0 = \frac{\partial \theta}{\partial \bar{z}}(0)$. Similarly, the convergence of $z\frac{\partial g_\varepsilon}{\partial z}$ implies

$$z\frac{\partial}{\partial z} \left(\frac{\theta_\varepsilon(z)}{z}\right) = z\left(\frac{1}{z} \frac{\partial \theta_\varepsilon}{\partial z}(z) - \frac{1}{z^2}\theta_\varepsilon(z)\right) = \frac{\partial \theta_\varepsilon}{\partial z}(z) - \frac{\theta_\varepsilon(z)}{z} \to 0,$$

and, since $\theta_\varepsilon(z)/z = g_\varepsilon(z) \to 1$ uniformly, it follows that $\frac{\partial \theta_\varepsilon}{\partial z}(z) = \frac{\partial \theta}{\partial z}(\varepsilon z)$ converges as $\varepsilon \to 0$ to $1 = \frac{\partial \theta}{\partial z}(0)$.

Having established that θ is a C^1-smooth function, we now take a closer look at the relation

$$\Psi_1(\theta, \eta) = \left(\theta^k + \check{X}_\eta(\theta^k, \hat{u} \circ \theta), \hat{u} \circ \theta + \hat{X}_\eta(\theta^k, \hat{u} \circ \theta)\right)$$

$$= (z^k, \hat{v}) = v_1, \tag{B.20}$$

viewed as a function of $z \in \dot{\mathbb{D}}_\rho$. Since $\theta(0) = 0$ and $d\theta(0) = \mathbb{1}$, the argument of Lemma B.29 implies that both $\hat{v}(z)$ and $\hat{u} \circ \theta(z)$ are $O(|z|^{k+1})$, so this equation implies

146 *Local Positivity of Intersections*

$$\widehat{X}_{\eta(z)}\left([\theta(z)]^k, \widehat{u} \circ \theta(z)\right) = O(|z|^{k+1}).$$

The second estimate in Lemma B.32 then gives

$$\left|\widehat{X}_{\eta(z)}\left([\theta(z)]^k, \widehat{u} \circ \theta(z)\right) - \eta(z)\right| \leq C|\widehat{u} \circ \theta(z)| \cdot |\eta(z)| \leq C'|z|^{k+1}$$

for some constant $C' > 0$. We conclude that η extends to a continuous function on $\mathbb{D}_\rho$ with $\eta(0) = 0$ and $\eta(z) = O(|z|^{k+1})$, and since $k + 1 \geq 2$, the latter implies that η is also differentiable at $z = 0$ with $d\eta(0) = 0$.

Finally, differentiating (B.20) at $z \neq 0$ gives

$$dv_1(z) = d_1\Psi_1(\theta(z), \eta(z)) \circ d\theta(z) + d_2\Psi_1(\theta(z), \eta(z)) \circ d\eta(z).$$

If $k \geq 2$, then in the limit as $z \to 0$, the first differential of Ψ_1 becomes $d_1\Psi_1(0,0) = d(u \circ \varphi)(0) = 0$, while the second becomes $d_2\Psi_1(0,0) = (0, \mathbb{1})$ since $\Psi_1(0,w) = (0,w)$, and since $dv_1(0)$ also vanishes, this proves $\lim_{z \to 0} d\eta(z) = 0$. The case $k = 1$ is slightly different since $dv_1(0)$ and $d_1\Psi_1(0,0) \circ d\theta(0) = d_1\Psi_1(0,0) = d(u \circ \varphi)(0)$ do not vanish, but instead they are identical, so we obtain the same conclusion about $d\eta(z)$. $\qquad\square$

B.2.5 Conclusion of the Proof

We cannot apply Proposition B.28 directly to the relation $v \circ \psi(z) = \Psi(\varphi \circ \theta(z), \eta(z))$ because $v \circ \psi$ is not a smooth map. However, we can write

$$\widetilde{\varphi} := \varphi \circ \theta \circ \psi^{-1} \quad \text{and} \quad \widetilde{\eta} := \eta \circ \psi^{-1},$$

and then apply the proposition to the relation

$$v(z) = \Psi(\widetilde{\varphi}(z), \widetilde{\eta}(z)).$$

Since $\widetilde{\eta}$ and $\widetilde{\varphi}$ are of class C^1, it follows that $\widetilde{\eta}$ is a solution to a linear Cauchy–Riemann type equation of class C^0, and the similarity principle (Corollary B.21) then implies that $\widetilde{\eta}$ is either identically zero near $z = 0$ or satisfies

$$\widetilde{\eta}(z) = z^\ell A + o(|z|^\ell)$$

for some $\ell \in \mathbb{N}$ and $A \in \mathbb{C}^{n-1}\setminus\{0\}$. If $\widetilde{\eta}$ vanishes near 0, then so does η, and we obtain

$$(z^k, \widehat{v}(z)) = v \circ \psi(z) = \Psi(\varphi \circ \theta(z), 0) = (u \circ \varphi)(\theta(z)) = \left([\theta(z)]^k, \widehat{u} \circ \theta(z)\right).$$

Given that θ is of class C^1 with $d\theta(0) = \mathbb{1}$, this can only hold if θ is the identity map near $z = 0$, implying $\widehat{u} \equiv \widehat{v}$.

B.2 The Representation Formula

If, on the other hand, $\overline{\eta}(z) = z^\ell A + |z|^\ell R(z)$ with $A \neq 0$ and $\lim_{z \to 0} R(z) = 0$, then since $\psi(0) = 0$ and $d\psi(0) = \mathbb{1}$, we can write $\psi(z) = z + |z| \cdot r(z)$ with $\lim_{z \to 0} r(z) = 0$ and find

$$\eta(z) = \overline{\eta}(\psi(z)) = (z + |z| r(z))^\ell A + \left| z + |z| \cdot r(z) \right|^\ell R(z + |z| \cdot r(z)) = z^\ell A + o(|z|^\ell).$$

Since $\eta(z) = O(|z|^{k+1})$ by Lemma B.31, we deduce from this that $\ell > k$. It remains to relate this to the function $h(z) = (0, \widehat{h}(z)) := (0, \widehat{v}(z) - \widehat{u}(z)) = v \circ \psi(z) - u \circ \varphi(z)$, which can now be expressed as

$$\begin{aligned}
(0, \widehat{h}) &= \Psi(\varphi \circ \theta, \eta) - \Psi(\varphi, 0) = (u \circ \varphi) \circ \theta - u \circ \varphi + X_\eta(u \circ \varphi \circ \theta) \\
&= (\theta^k, \widehat{u} \circ \theta) - (z^k, \widehat{u}) + X_\eta(\theta^k, \widehat{u} \circ \theta) \\
&= \left(\theta^k - z^k + \check{X}_\eta(\theta^k, \widehat{u} \circ \theta), \widehat{u} \circ \theta - \widehat{u} + \widehat{X}_\eta(\theta^k, \widehat{u} \circ \theta) \right). \tag{B.21}
\end{aligned}$$

Since $\eta(z) = O(|z|^\ell)$ and $\widehat{u} \circ \theta(z) = O(|z|^{k+1})$, Lemma B.32 implies an estimate

$$\left| \check{X}_{\eta(z)}([\theta(z)]^k, \widehat{u} \circ \theta(z)) \right| \leq C|\eta(z)| \cdot |\widehat{u} \circ \theta(z)| = O(|z|^{\ell+k+1}),$$

so that (B.21) then gives $[\theta(z)]^k - z^k = O(|z|^{\ell+k+1})$. Since $\theta(z)/z$ can be assumed arbitrarily close to 1 for $|z|$ small, we then have

$$\begin{aligned}
|\theta - z| &= \left| \frac{\theta^k - z^k}{\theta^{k-1} + \theta^{k-2} z + \cdots + \theta z^{k-2} + z^{k-1}} \right| \\
&= \frac{|\theta^k - z^k|}{|z|^{k-1}} \frac{1}{\left| \left(\frac{\theta}{z}\right)^{k-1} + \cdots + \left(\frac{\theta}{z}\right) + 1 \right|} \\
&\leq \text{const} \cdot \frac{|z|^{\ell+k+1}}{|z|^{k-1}} = O(|z|^{\ell+2}),
\end{aligned}$$

which implies an estimate of the form

$$|\widehat{u}(\theta(z)) - \widehat{u}(z)| \leq C|\theta(z) - z| = O(|z|^{\ell+2}) \tag{B.22}$$

since $\widehat{u}$ is of class C^1. Finally, the second estimate in Lemma B.32 implies

$$\left| \widehat{X}_{\eta(z)}([\theta(z)]^k, \widehat{u} \circ \theta(z)) - \eta(z) \right| \leq C|\eta(z)| \cdot |\widehat{u} \circ \theta(z)| = O(|z|^{\ell+k+1});$$

hence

$$\begin{aligned}
\widehat{X}_{\eta(z)}([\theta(z)]^k, \widehat{u} \circ \theta(z)) &= \eta(z) + O(|z|^{\ell+k+1}) = z^\ell A + o(|z|^\ell) + O(|z|^{\ell+k+1}) \\
&= z^\ell A + o(|z|^\ell),
\end{aligned}$$

and combining this with (B.22), we can now derive from (B.21) the relation

$$\widehat{h}(z) = O(|z|^{\ell+2}) + z^\ell A + o(|z|^\ell) = z^\ell A + o(|z|^\ell).$$

The proof of Theorem B.23 is now complete.

148 *Local Positivity of Intersections*

B.3 Counting Local Intersections and Singularities

In this section, we take the local representation formula of Theorem B.23 as a black box and use it to deduce the standard results on positivity of intersections.

According to the representation formula, a nonconstant J-holomorphic curve has a well-defined tangent space at every point, including critical points, with a nonnegative *critical order* $k \in \mathbb{Z}$ that is strictly positive if and only if the point is critical. We can now prove local positivity of intersections (Theorem 2.3) by considering separately the cases where the two curves have matching or nonmatching tangent spaces at their intersection. Note that when $\dim M = 4$, the condition that two (complex-linear!) tangent spaces at an intersection point do not match means simply that they are transverse, and the intersection itself is then transverse if and only if neither curve is critical at the intersection point.

Exercise B.36 Let $\pi \colon \mathbb{C}^n \backslash \{0\} \to \mathbb{C}\mathbb{P}^{n-1}$ denote the natural projection, and consider a map $u \colon \mathbb{D} \to \mathbb{C}^n$ of the form $u(z) = (z^k, |z|^{k+1} f(z))$ for some $k \geq \mathbb{N}$ and a bounded function $f \colon \mathbb{D} \to \mathbb{C}^{n-1}$. Show that for any neighborhood $\mathcal{U}$ of $[1 : 0 : \ldots : 0] \in \mathbb{C}\mathbb{P}^{n-1}$, one can find $\rho > 0$ such that the restriction of $\pi \circ u$ to $\mathbb{D}_\rho \backslash \{0\}$ has image in $\mathcal{U}$.

Proposition B.37 *Suppose $u \colon (\Sigma, j) \to (M, J)$ and $v \colon (\Sigma', j') \to (M, J)$ are two J-holomorphic curves with an intersection $u(z_0) = v(\zeta_0)$ at which u has critical order $k_u - 1 \geq 0$, v has critical order $k_v - 1 \geq 0$, and their tangent spaces (in the sense of Theorem B.23) are distinct. Then the intersection is isolated, and, if $\dim M = 4$, its local intersection index is*

$$\iota(u, z_0 \, ; \, v, \zeta_0) = k_u k_v;$$

in particular, it is positive, and equal to 1 if and only if the intersection is transverse.

Proof By Theorem B.23, we can choose C^1-smooth coordinates such that without loss of generality $z_0 = \zeta_0 = 0 \in \mathbb{D} = \Sigma = \Sigma'$, $M = \mathbb{C}^n$, $u(z) = (z^{k_u}, |z|^{k_u+1} f(z))$ for some bounded function $f \colon \mathbb{D} \to \mathbb{C}^{n-1}$, and $v \colon \mathbb{D} \to \mathbb{C}^n$ satisfies $v(0) = 0$. The condition of distinct tangent spaces implies via Exercise B.36 that if $\pi \colon \mathbb{C}^n \backslash \{0\} \to \mathbb{C}\mathbb{P}^{n-1}$ denotes the natural projection, we can also assume that the images of the maps

$$\pi \circ u|_{\mathbb{D} \backslash \{0\}}, \ \pi \circ v|_{\mathbb{D} \backslash \{0\}} \colon \mathbb{D} \backslash \{0\} \to \mathbb{C}\mathbb{P}^{n-1}$$

lie in arbitrarily small neighborhoods of two distinct points. The same is also true if we replace u with any of the maps

$$u_\tau \colon \mathbb{D} \to \mathbb{C}^n \colon z \mapsto (z^{k_u}, \tau |z|^{k_u+1} f(z)), \qquad \tau \in [0, 1].$$

B.3 Counting Local Intersections and Singularities 149

The claim that the intersection is isolated follows immediately, and, when $n = 2$, we also deduce via Exercise 2.1 that $\iota(u, 0; v, 0) = \iota(u_0, 0; v, 0)$. After applying the same homotopy argument in different coordinates adapted to v and then choosing new coordinates so that the tangent spaces of u and v match $\mathbb{C} \times \{0\}$ and $\{0\} \times \mathbb{C}$, respectively, we can reduce the problem to a computation of $\iota(u_0, 0; v_0, 0)$ for

$$u_0(z) = (z^{k_u}, 0), \qquad v_0(z) = (0, z^{k_v}).$$

Choose $\epsilon \in \mathbb{C} \backslash \{0\}$ and perturb these maps to $(z^{k_u} + \epsilon, 0)$ and $(0, z^{k_v} + \epsilon)$, respectively. Both are now holomorphic for the standard complex structure on $\mathbb{C}^2$ and they have exactly $k_u k_v$ intersections, all transverse. $\qquad \square$

When both curves have matching tangent spaces where they intersect, we will need to use the more precise information provided by Theorem B.23. Observe that in this case the intersection can never be transverse.

Exercise B.38 Suppose $\dim M = 4$, $u, v \colon (\mathbb{D}, i) \to (M, J)$ are J-holomorphic disks and they have an isolated intersection $u(0) = v(0)$. Given $k, \ell \in \mathbb{N}$, define the J-holomorphic branched covers $u^k, v^\ell \colon (\mathbb{D}, i) \to (M, J)$,

$$u^k(z) := u(z^k), \qquad v^\ell(z) := v(z^\ell).$$

Show that $\iota(u^k, 0; v^\ell, 0) = k\ell \cdot \iota(u, 0; v, 0)$.

Proposition B.39 *Suppose $u \colon (\Sigma, j) \to (M, J)$ and $v \colon (\Sigma', j') \to (M, J)$ are two J-holomorphic curves with an intersection $u(z_0) = v(\zeta_0)$ at which u has critical order $k_u - 1 \geq 0$, v has critical order $k_v - 1 \geq 0$, and their tangent spaces (in the sense of Theorem B.23) are identical. Then either the intersection $u(z_0) = v(\zeta_0)$ is isolated or there exist neighborhoods $z_0 \in \mathcal{U}_{z_0} \subset \Sigma$ and $\zeta_0 \in \mathcal{U}_{\zeta_0} \subset \Sigma'$ such that $u(\mathcal{U}_{z_0}) = v(\mathcal{U}_{\zeta_0})$. In the former case, if $\dim M = 4$, the local intersection index satisfies*

$$\iota(u, z_0; v, \zeta_0) > k_u k_v;$$

in particular, it is strictly greater than 1.

Proof We can choose holomorphic coordinates near $z_0 \in \Sigma$ and $\zeta_0 \in \Sigma'$ so that, without loss of generality, $(\Sigma, j) = (\Sigma', j') = (\mathbb{D}, i)$ with $z_0 = \zeta_0 = 0$. Since k_u and k_v may be different, we first replace u and v with suitable branched covers so that their critical orders become the same: let

$$m = k_u k_v \in \mathbb{N},$$

and define $u', v' \colon (\mathbb{D}, i) \to (M, J)$ by

$$u'(z) := u(z^{k_v}), \qquad v'(z) := v(z^{k_u}),$$

150 *Local Positivity of Intersections*

so that, in particular, u' and v' both have critical order $m - 1$ at the intersection $u'(0) = v'(0)$, as well as matching tangent spaces. Now by Theorem B.23, we find new choices of C^1-smooth local coordinates in $\mathbb{D}$ near 0 and smooth coordinates in M near $u(0) = v(0)$ such that

$$u'(z) = (z^m, \widehat{u}(z)), \qquad v'(z) = (z^m, \widehat{v}(z))$$

for some functions $\widehat{u}, \widehat{v} \colon \mathbb{D} \to \mathbb{C}^{n-1}$ of class C^1 that are both $O(|z|^{m+1})$. For each $j = 0, \dots, m - 1$, we can also compose u' with the smooth coordinate change $z \mapsto e^{2\pi i j/m} z$ to produce a new parametrization $v'_j \colon \mathbb{D} \to \mathbb{C}^n$ of the form

$$v'_j(z) := v'(e^{2\pi i j/m} z) = (z^m, \widehat{v}_j(z)), \qquad \text{where} \quad \widehat{v}_j(z) = \widehat{v}(e^{2\pi i j/m} z),$$

for which the statement of Theorem B.23 is equally valid. If $\widehat{u} - \widehat{v}_j$ is identically zero for some $j = 0, \dots, m - 1$, then we have

$$u'(z) = v'(e^{2\pi i j/m} z) \quad \text{for all } z \in \mathbb{D},$$

implying that u' and v' have identical images on some neighborhood of the intersection, in which case so do u and v. If not, then Theorem B.23 gives for each $j = 0, \dots, m - 1$ the formula

$$\widehat{u}(z) - \widehat{v}_j(z) = z^{m+\ell_j} C_j + |z|^{m+\ell_j} r_j(z), \tag{B.23}$$

where $C_j \in \mathbb{C}^{n-1} \backslash \{0\}$, $\ell_j \in \mathbb{N}$ and $r_j(z) \in \mathbb{C}^{n-1}$ is a function with $r_j(z) \to 0$ as $z \to 0$. This expression has an isolated zero at $z = 0$, and thus the intersection of u' and v' (and hence of u and v) is isolated.

If $n = 2$, we can now compute $\iota(u', 0; v', 0)$ by choosing $\epsilon \in \mathbb{C} \backslash \{0\}$ small and defining the perturbation

$$u'_\epsilon(z) := (z^m, \widehat{u}(z) + \epsilon).$$

This curve does not intersect v' at $z = 0$ since $\epsilon \neq 0$. If $u'_\epsilon(z) = v'(\zeta)$, then $z^m = \zeta^m$; hence $\zeta = e^{2\pi i j/m} z$ for some $j = 0, \dots, m - 1$, and equality in the second factor then implies

$$\widehat{v}_j(z) - \widehat{u}(z) = \epsilon. \tag{B.24}$$

By (B.23), the zero of $\widehat{v}_j(z) - \widehat{u}(z)$ at $z = 0$ has order $m + \ell_j > m$; thus, if $\epsilon \in \mathbb{C}$ is sufficiently small and chosen generically so that it is a regular value of $\widehat{v}_j - \widehat{u}$, we conclude that (B.24) has exactly $m + \ell_j$ solutions near $z = 0$, all of them simple positive zeroes of $\widehat{v}_j - \widehat{u} - \epsilon$ and thus corresponding to transverse positive intersections of u'_ϵ with v'. Adding these up for all choices of $j = 0, \dots, m - 1$, we conclude

$$\iota(u', 0; v', 0) > m^2 = k_u^2 k_v^2,$$

so by Exercise B.38, $\iota(u, 0; v, 0) > k_u k_v$. $\qquad\qquad\square$

B.3 Counting Local Intersections and Singularities 151

Exercise B.40 Find examples to show that in the situation of Proposition B.39, $\iota(u, z_0 ; v, \zeta_0)$ cannot in general be bounded from above.

Combining Propositions B.37 and B.39 completes the proof of Theorem 2.3.

We now turn to the proof of Lemma 2.6 from Lecture 2, which asserts that any critical point on a simple J-holomorphic curve gives rise to a strictly positive count of double points after an immersed perturbation. In the background of this statement is the fact that all simple holomorphic curves are locally injective, which we can now prove using the representation formula of Theorem B.23.

Proposition B.41 *Suppose* $u \colon (\Sigma, j) \to (M, J)$ *is a J-holomorphic curve that is nonconstant near a point* $z_0 \in \Sigma$ *with* $du(z_0) = 0$. *Then there exists a neighborhood* $z_0 \in \mathcal{U}_{z_0} \subset \Sigma$ *such that there is a biholomorphic identification*

$$\varphi \colon (\mathbb{D}, i) \xrightarrow{\cong} (\mathcal{U}_{z_0}, j)$$

with $\varphi(0) = z_0$, *a number* $k \in \mathbb{N}$, *and an injective J-holomorphic map*

$$v \colon (\mathbb{D}, i) \to (M, J)$$

with

$$dv(z) \neq 0 \text{ for } z \in \mathbb{D} \setminus \{0\} \text{ and } u \circ \varphi(z) = v(z^k) \text{ for } z \in \mathbb{D}.$$

If $u \colon (\Sigma, j) \to (M, J)$ *is a simple curve, then* $k = 1$.

Proof Theorem B.23 provides C^1-smooth local coordinates near $z_0 \in \Sigma$ and smooth coordinates near $u(z_0) \in M$ in which u takes the form

$$u(z) = (z^k, \widehat{u}(z)) \in \mathbb{C}^n$$

for a C^1-smooth map $\widehat{u} \colon \mathbb{D} \to \mathbb{C}^{n-1}$ with $\widehat{u}(z) = O(|z|^{k+1})$, where $k - 1 \geq 0$ is the critical order of u at z_0, and all the maps in this picture are of class C^∞ away from $z_0 \in \Sigma$ or $0 \in \mathbb{D}$, respectively. For each $j = 1, \ldots, k - 1$, we can compose this representation of u with the smooth reparametrization $\psi_j(z) := e^{2\pi i j/k} z$ and thus use Theorem B.23 to compare u with

$$u_j(z) := u(e^{2\pi i j/k} z) = (z^k, \widehat{u}_j(z)), \quad \text{where} \quad \widehat{u}_j(z) := \widehat{u}(e^{2\pi i j/k} z).$$

The theorem implies that each $\widehat{u} - \widehat{u}_j$ is either identically zero or has an isolated zero at $z = 0$. Self-intersections $u(z) = u(\zeta)$ with $z \neq \zeta$ can now be identified with pairs $j \in \{1, \ldots, k-1\}$ and $z \in \mathbb{D}$ for which $\widehat{u}(z) = \widehat{u}_j(z)$. Let $m \in \{1, \ldots, k\}$ denote the smallest number for which $\widehat{u} \equiv \widehat{u}_m$; hence $u(z) = u(e^{2\pi i m/k} z)$ for all z. Then we also have $\widehat{u} \equiv \widehat{u}_{jm}$ for all $j \in \mathbb{Z}$, so m must divide k, and, setting

152 *Local Positivity of Intersections*

$\ell := k/m$, we see that $u \colon \mathbb{D} \to \mathbb{C}^n$ is invariant with respect to the $\mathbb{Z}_\ell$ action on $\mathbb{D}$ generated by the rotation $\psi := \psi_m$. It therefore factors as

$$u(z) = v(z^\ell)$$

for a continuous map $v \colon \mathbb{D} \to \mathbb{C}^n$ that is smooth on $\mathbb{D} \backslash \{0\}$, and v is injective near 0 since we always have $\widehat{u}(z) \neq \widehat{u}_j(z)$ near $z = 0$ for $j = 1, \ldots, m-1$.

It remains to show that $v \colon \mathbb{D} \to \mathbb{C}^n$ can be reparametrized near $0 \in \mathbb{D}$ to become a *smooth J*-holomorphic curve. We shall deduce this from elliptic regularity, but first, we need to switch back to smooth holomorphic coordinates on the domain. Since the parametrization $u(z) = (z^k, \widehat{u}(z))$ was obtained via a C^1-smooth coordinate chart on the smooth Riemann surface (Σ, j), this parametrization is a pseudoholomorphic map $(\mathbb{D}, j') \to (M, J)$ for a continuous complex structure j' that is smooth on $\dot{\mathbb{D}} := \mathbb{D} \backslash \{0\}$ and uniquely determined there by $j' = u^*J$. It follows that the $\mathbb{Z}_\ell$-action on $\mathbb{D}$ leaving u invariant acts holomorphically on $(\mathbb{D}, j')$, and it can therefore be defined as a group of biholomorphic (and therefore smooth) transformations on the simply connected neighborhood $\mathcal{U} \subset \Sigma$ of z_0 that is identified with $\mathbb{D}$ via our C^1-coordinates. Using the Riemann mapping theorem, we can now choose a holomorphic coordinate chart identifying $(\mathcal{U}, j)$ with $(\mathbb{D}, i)$ and z_0 with $0 \in \mathbb{D}$, so that in the new coordinates, ψ generates a $\mathbb{Z}_\ell$-action by biholomorphic transformations on $(\mathbb{D}, i)$ that fix 0. All such transformations are rotations, and thus ψ is given by the same formula as before in the new coordinates, and we can define a continuous map $v \colon \mathbb{D} \to \mathbb{C}^n$ as before via the relation $u(z) = v(z^\ell)$, observing that v is manifestly smooth and holomorphic on the standard punctured disk $(\dot{\mathbb{D}}, i)$. Since $du(z) = O(|z|^{k-1})$, we then deduce from $u(z) = v(z^\ell)$ and $du(z) = dv(z^\ell) \circ (\ell z^{\ell-1})$ an estimate of the form

$$\left| dv(z^\ell) \right| \leq C \frac{|du(z)|}{|z|^{\ell-1}} \leq C' |z|^{k-\ell}$$

near $z = 0$. This expression is bounded since $\ell \leq k$, implying via Exercise B.16 that the map $v \colon \mathbb{D} \to \mathbb{C}^n$ is of class $W^{1,\infty}$. It is therefore smooth by Proposition B.15. $\qquad\square$

The remainder of Lemma 2.6 can be restated as follows:

Proposition B.42 *Suppose* dim $M = 4$ *and* $u \colon (\mathbb{D}, i) \to (M, J)$ *is an injective J-holomorphic map with critical order* $k - 1 \geq 1$ *at* $z = 0$ *and no critical points on* $\mathbb{D} \backslash \{0\}$. *Then there exists an integer*

$$\delta(u, 0) \geq \frac{k(k-1)}{2}$$

B.3 Counting Local Intersections and Singularities 153

depending only on the germ of u near 0 such that for any given neighborhood $\mathcal{U} \subset \mathbb{D}$ of 0 and symplectic form ω_0 defined near $u(0)$ taming J, one can find a C^1-smooth map $u_\epsilon \colon \mathbb{D} \to M$ satisfying the following conditions:

(1) *u_ϵ is C^1-close to u and matches u outside $\mathcal{U}$ and at 0.*
(2) *u_ϵ is an immersion with $u_\epsilon^* \omega_0 > 0$.*
(3) *u_ϵ has finitely many self-intersections and satisfies*

$$\frac{1}{2} \sum_{(z,\zeta)} \iota(u_\epsilon, z \,;\, u_\epsilon, \zeta) = \delta(u, 0), \tag{B.25}$$

where the sum ranges over all pairs $(z, \zeta) \in \mathbb{D} \times \mathbb{D}$ such that $z \neq \zeta$ and $u_\epsilon(z) = u_\epsilon(\zeta)$.

Our proof will show, in fact, that the tangent spaces spanned by the perturbation u_ϵ can be arranged to be uniformly close to i-complex subspaces (or equivalently J-complex subspaces, since J and i may also be assumed uniformly close in a small enough neighborhood of $u(0)$). This implies that it is a symplectic immersion without loss of generality for any given ω_0 taming J, as the condition of being a symplectic subspace is open. In practice, the crucial point in applications is that the complex structure on the bundle $(u_\epsilon^* TM, J)$ admits a homotopy supported near 0 to a new complex structure for which $\operatorname{im} du_\epsilon$ becomes a complex subbundle – in this way we can keep control over the c_1 term in the adjunction formula. The subtlety in the proof is that the change in tangent subspaces when perturbing from u to u_ϵ cannot be understood as a C^0-small perturbation if $du(0) = 0$. Our strategy will be to show that the tangent spaces spanned by du_ϵ are, in fact, C^0-close to the tangent spaces spanned by *another* map that is a holomorphic immersion. In order to make this notion precise, we need a practical way of measuring the "distance" between two subspaces of a vector space, in particular for the case when both subspaces arise as images of injective linear maps.

Definition B.43 Fix the standard Euclidean norm on $\mathbb{R}^n$. Given two subspaces $V, W \subset \mathbb{R}^n$ of the same positive dimension, define

$$\operatorname{dist}(V, W) := \max_{v \in V, |v|=1} \operatorname{dist}(v, W) := \max_{v \in V, |v|=1} \min_{w \in W} |v - w|.$$

Definition B.44 The **injectivity modulus** of a linear map $A \colon \mathbb{R}^k \to \mathbb{R}^n$ is

$$\operatorname{Inj}(A) = \min_{v \in \mathbb{R}^k \setminus \{0\}} \frac{|Av|}{|v|} \geq 0.$$

Clearly, $\operatorname{Inj}(A) > 0$ if and only if A is injective.

154   *Local Positivity of Intersections*

Lemma B.45 *For any pair of injective linear maps $A, B \colon \mathbb{R}^k \to \mathbb{R}^n$,*

$$\operatorname{dist}(\operatorname{im} A, \operatorname{im} B) \le \frac{|A - B|}{\operatorname{Inj}(A)}.$$

Proof Pick any nonzero vector $v \in \mathbb{R}^n$. Then $Av \ne 0$ since A is injective, and we have

$$\operatorname{dist}\left(\frac{Av}{|Av|}, \operatorname{im} B\right) = \min_{w \in \mathbb{R}^k} \left|A\frac{v}{|Av|} - Bw\right| \le \left|A\frac{v}{|Av|} - B\frac{v}{|Av|}\right|$$

$$\le |A - B|\frac{|v|}{|Av|} \le \frac{|A - B|}{\operatorname{Inj}(A)}. \qquad\qquad \square$$

Lemma B.46 *Given a symplectic form ω_0 on $\mathbb{C}^2$ taming i, there exists $\epsilon > 0$ such that if $V \subset \mathbb{C}^2$ is a complex 1-dimensional subspace, then all real 2-dimensional subspaces $W \subset \mathbb{C}^2$ satisfying $\operatorname{dist}(V, W) < \epsilon$ are ω_0-symplectic.*

Exercise B.47 Prove the lemma. *Hint:* $\mathbb{CP}^1$ is compact.

Proof of Proposition B.42 By Theorem B.23, we can assume after choosing suitable C^1-smooth coordinates near $0 \in \mathbb{D}$ and smooth coordinates near $u(0) \in M$ that

$$u(z) = (z^k, \hat{u}(z)) \in \mathbb{C}^2$$

for some integer $k \ge 2$, where the almost complex structure J matches i at $0 \in \mathbb{C}^2$ and $\hat{u}$ is a map $\mathbb{D}_\rho \to \mathbb{C}$ of class C^1 on a disk of some radius $\rho > 0$ such that the other branches

$$u_j(z) := u(e^{2\pi i j/k}z) = (z^k, \hat{u}_j(z)), \qquad \hat{u}_j(z) := \hat{u}(e^{2\pi i j/k}z),$$

for $j = 1, \ldots, k - 1$ are related by

$$\hat{u}_j(z) - \hat{u}(z) = z^{k+\ell_j}C_j + |z|^{k+\ell_j}r_j(z) \tag{B.26}$$

for some $\ell_j \in \mathbb{N}$, $C_j \in \mathbb{C}\backslash\{0\}$ and $r_j \colon \mathbb{D}_\rho \to \mathbb{C}$ with $r_j(z) \to 0$ as $z \to 0$. Here we've used the assumption that u is injective in order to conclude that $\hat{u}_j - \hat{u}$ is not identically zero. By shrinking $\rho > 0$ if necessary, we can also assume u is embedded on $\mathbb{D}_\rho\backslash\{0\}$, and that the symplectic form ω_0, which tames J by assumption, also tames i on some neighborhood of $u(\mathbb{D}_\rho)$. Fix a smooth cutoff function $\beta \colon \mathbb{D}_\rho \to [0, 1]$ that equals 1 on $\mathbb{D}_{\rho/2}$ and has compact support in the interior. Then for $\epsilon \in \mathbb{C}$ sufficiently close to 0, consider the C^1-close perturbation

$$u_\epsilon(z) := (z^k, \hat{u}(z) + \epsilon\beta(z)z),$$

B.3 Counting Local Intersections and Singularities 155

which satisfies $u_\epsilon(0) = 0$ and is immersed if $\epsilon \neq 0$. Since u is embedded on $\mathbb{D}_\rho \backslash \mathbb{D}_{\rho/2}$, we may assume for $|\epsilon|$ sufficiently small that u_ϵ has no self-intersections outside of the region where $\beta \equiv 1$. Then a self-intersection $u_\epsilon(z) = u_\epsilon(\zeta)$ with $z \neq \zeta$ occurs wherever $\zeta = e^{2\pi ij/k}z \neq 0$ for some $j = 1, \ldots, k-1$ and $\hat{u}(z) + \epsilon z = \hat{u}_j(z) + \epsilon e^{2\pi ij/k}z$, which by (B.26) means

$$z^{k+\ell_j}C_j + |z|^{k+\ell_j}r_j(z) + \epsilon\left(e^{2\pi ij/k} - 1\right)z = 0.$$

Assume $\epsilon \in \mathbb{C} \backslash \{0\}$ is chosen generically so that the zeroes of this function are all simple (see Exercise B.49). Then each zero other than the "trivial" solution at $z = 0$ represents a transverse (positive or negative) self-intersection of u_ϵ, and the algebraic count of these (discounting the trivial solution) for $|\epsilon|$ sufficiently small is $k + \ell_j - 1 \geq k$. Adding these up for all $j = 1, \ldots, k-1$, we obtain

$$\delta(u, 0) := \frac{1}{2}\sum_{(z,\zeta)}\iota(u_\epsilon, z; u_\epsilon, \zeta) = \frac{1}{2}\sum_{j=1}^{k-1}(k + \ell_j - 1) \geq \frac{1}{2}k(k-1). \tag{B.27}$$

It remains to show that u_ϵ satisfies $u_\epsilon^*\omega_0 > 0$, which is equivalent to showing that im $du_\epsilon(z) \subset \mathbb{C}^2$ is an ω_0-symplectic subspace for all z. By Theorem B.23, there exist constants $\ell \in \mathbb{N}$ and $C \in \mathbb{C} \backslash \{0\}$ such that

$$\hat{u}(z) = z^{k+\ell}C + o(|z|^{k+\ell}), \tag{B.28}$$

and we claim that the formula

$$d\hat{u}(z) = (k + \ell)z^{k+\ell-1}C + o(|z|^{k+\ell-1}) \tag{B.29}$$

also holds. If $\hat{u}$ were smooth, this would follow immediately from (B.28) via Taylor's theorem, but we have to work a little bit harder since $\hat{u}$ is only of class C^1. Recall that $\hat{u}$ is a composition of the form $\hat{u} = u_2 \circ \varphi$, where we can take $\varphi: \mathbb{D}_\rho \to \mathbb{D}$ to be a C^1-smooth local diffeomorphism with $\varphi(0) = 0$ and $d\varphi(0) = 1$, and $u_2: \mathbb{D}_\epsilon \to \mathbb{C}$ is a map of class C^∞, i.e., the second coordinate of the original J-holomorphic curve before it was nonsmoothly reparametrized. Since $\varphi(z) = z + o(|z|)$ and $\varphi^{-1}(z) = z + o(|z|)$, we can write $\varphi^{-1}(z) = z + |z| \cdot r(z)$ with $\lim_{z\to 0} r(z) = 0$ and write (B.28) as $\hat{u}(z) = z^{k+\ell}C + |z|^{k+\ell}R(z)$ with $\lim_{z\to 0} R(z) = 0$, implying

$$u_2(z) = \hat{u}(z + |z| \cdot r(z)) = (z + |z| \cdot r(z))^{k+\ell}C$$
$$+ |z + |z| \cdot r(z)|^{k+\ell}R(z + |z| \cdot r(z)) = z^{k+\ell}C + o(|z|^{k+\ell}).$$

Now since u_2 is smooth, this expression can be interpreted as saying that $z^{k+\ell}C$ is the lowest-order nontrivial term in its Taylor series, and we can then draw a

156 Local Positivity of Intersections

similar conclusion for du_2: namely for $C' := (k + \ell)C$ and a function $R'(z) \in \mathbb{C}$ with $\lim_{z \to 0} R'(z) = 0$, we have

$$du_2(z) = z^{k+\ell-1} C' + o(|z|^{k+\ell-1}) = z^{k+\ell-1} C' + |z|^{k+\ell-1} R'(z).$$

Finally, reverse the process: writing $\varphi(z) = z + |z| \cdot r'(z)$ with $\lim_{z \to 0} r'(z) = 0$, the relation (B.29) follows from

$$
\begin{aligned}
d\hat{u}(z) &= du_2(z + |z| \cdot r'(z)) \circ d\varphi(z) \\
&= \left[(z + |z| \cdot r'(z))^{k+\ell-1} C' + \left| z + |z| \cdot r'(z) \right|^{k+\ell-1} R'(z) \right] \circ d\varphi(z) \\
&= \left[(z + |z| \cdot r'(z))^{k+\ell-1} C' + \left| z + |z| \cdot r'(z) \right|^{k+\ell-1} R'(z) \right] \\
&\quad + \left[(z + |z| \cdot r'(z))^{k+\ell-1} C' + \left| z + |z| \cdot r'(z) \right|^{k+\ell-1} R'(z) \right] \\
&\quad \circ (d\varphi(z) - d\varphi(0)) \\
&= z^{k+\ell-1} C' + o(|z|^{k+\ell-1}),
\end{aligned}
$$

where the existence of a suitable remainder function depends on the fact that $d\varphi(z) - d\varphi(0)$ is a continuous function of z that vanishes at $z = 0$.

We would now like to compare u_ϵ with the holomorphic polynomial

$$P_\epsilon: \mathbb{D}_\rho \to \mathbb{C}^2 : z \mapsto (z^k, z^{k+\ell} C + \epsilon z),$$

which, due to (B.29), satisfies

$$du_\epsilon(z) - dP_\epsilon(z) = |z|^{k+\ell-1} R(z)$$

for a remainder term $R(z) \in \mathbb{C}^2$ that satisfies $\lim_{z \to 0} R(z) = 0$ and does not depend on ϵ. Abbreviating $A_\epsilon(z) := dP_\epsilon(z)$ and $B_\epsilon(z) := du_\epsilon(z)$, this gives an estimate of the form

$$|A_\epsilon(z) - B_\epsilon(z)| \le c_1 |z|^{k+\ell-1}$$

for some constant $c_1 > 0$ independent of ϵ. Computing $dP_\epsilon(0)$, we find similarly a constant $c_2 > 0$ independent of ϵ such that

$$|A_\epsilon(z)v| \ge c_2 |z|^{k-1} |v| \quad \text{for all } v \in \mathbb{C};$$

thus $\mathrm{Inj}(A_\epsilon(z)) \ge c_2 |z|^{k-1}$, and

$$\frac{|A_\epsilon(z) - B_\epsilon(z)|}{\mathrm{Inj}(A_\epsilon(z))} \le c_3 |z|^\ell$$

for some constant $c_3 > 0$ independent of ϵ. Now since P_ϵ is holomorphic (for the standard complex structure) for all ϵ, $\mathrm{im}\, A_\epsilon(z) \subset \mathbb{C}^2$ is always complex linear, so the above estimates imply together with Lemmas B.45 and B.46 that for a sufficiently small radius $\rho_0 > 0$, the images of $du_\epsilon(z)$ for all $z \in \mathbb{D}_{\rho_0} \setminus \{0\}$

B.3 Counting Local Intersections and Singularities 157

and $\epsilon \in \mathbb{D}_{\rho_0}$ are ω_0-symplectic. This is also true for $z = 0$ if $\epsilon \neq 0$, since then $du_\epsilon(0) = dP_\epsilon(0)$ is complex linear.

To conclude, fix $\rho_0 > 0$ as above and choose $\epsilon \in \mathbb{C}\backslash\{0\}$ sufficiently close to 0 so that outside of $\mathbb{D}_{\rho_0}$, u_ϵ is C^1-close enough to u for its tangent spaces to be ω_0-symplectic (recall that J is also ω_0-tame). The previous paragraph then implies that the tangent spaces of u_ϵ are ω_0-symplectic everywhere. $\qquad\square$

Exercise B.48 Verify that the formula obtained in (B.27) for $\delta(u, 0)$ does not depend on any choices.

Exercise B.49 Assume $f : \mathcal{U} \to \mathbb{C}$ is a C^1-smooth map on a domain $\mathcal{U} \subset \mathbb{C}$ containing 0, with $f(0) = 0$ and $df(0) = 0$. Show that for almost every $\epsilon \in \mathbb{C}$, the map $f_\epsilon : \mathcal{U} \to \mathbb{C} : z \mapsto f(z) + \epsilon z$ has 0 as a regular value. *Hint:* Use the implicit function theorem to show that the set

$$X := \{(\epsilon, z) \in \mathbb{C} \times (\mathcal{U}\backslash\{0\}) \mid f_\epsilon(z) = 0\}$$

is a smooth submanifold of $\mathbb{C}^2$, and a point $(\epsilon, z) \in X$ is regular for the projection $\pi : X \to \mathbb{C} : (\epsilon, z) \mapsto \epsilon$ if and only if z is a regular point of f_ϵ. Then apply Sard's theorem to π.[1]

Exercise B.50 Find examples to show that the bound $\delta(u, 0) \geq \frac{k(k-1)}{2}$ in Proposition B.42 is sharp and that there is no similar upper bound for $\delta(u, 0)$ in terms of k (*cf. Exercise B.40*).

---

[1] Note that while Sard's theorem is often stated only for C^∞-smooth maps, it is valid more generally for continuously differentiable maps $f : M \to N$ of class C^{m-n+1} for $m := \dim M$ and $n := \dim N$ (see [Sar42]).

Appendix C

A Quick Survey of Siefring's Intersection Theory

This appendix is meant in part as a survey and also as a quick reference guide for the intersection theory of punctured holomorphic curves. Except where otherwise noted, the proofs of everything stated below are due to Siefring [Sie11], and the details (modulo proofs of the relative asymptotic formulas) can be found in Lectures 3 and 4 of this book. Since intersection theory has also played a large role in the development of Hutchings's embedded contact homology (ECH), we will simultaneously take the opportunity to clarify some of the connections between Siefring's theory and equivalent notions that often appear (sometimes with very different notation) in the ECH literature. For an important word of caution about notational differences between this book and [Sie11], see Remark 4.7.

C.1 Preliminaries

Assume M is a closed oriented 3-manifold with a stable Hamiltonian structure (ω, λ), i.e., a 2-form ω and 1-form λ that satisfy $d\omega = 0$, $\lambda \wedge \omega > 0$ and $\ker \omega \subset \ker d\lambda$. (The reader unfamiliar with or uninterested in stable Hamiltonian structures is free to assume $(\omega, \lambda) = (d\alpha, \alpha)$ where α is a contact form.) This data determines an oriented 2-plane field

$$\xi = \ker \lambda \subset TM$$

and a **Reeb vector field** R such that

$$\omega(R, \cdot) \equiv 0 \quad \text{and} \quad \lambda(R) \equiv 1.$$

We assume throughout the following that all closed orbits of R are nondegenerate. As mentioned in the footnote to Theorem 4.1, the major results continue to hold without serious changes if orbits are Morse–Bott, as long as

C.1 Preliminaries

homotopies of asymptotically cylindrical maps are required to fix the asymptotic orbits in place. There also exists a generalization of the theory that lifts the latter condition (see [Wen10a, §4.1], and [SW]).

Suppose γ is a closed orbit of R and τ is a choice of trivialization of ξ along γ. The Conley–Zehnder index of γ relative to this trivialization will be denoted by

$$\mu_{CZ}^\tau(\gamma) \in \mathbb{Z}.$$

If γ has period $T > 0$, then any choice of ω-compatible complex structure J on ξ and parametrization $\gamma \colon S^1 := \mathbb{R}/\mathbb{Z} \to M$ satisfying $\lambda(\dot\gamma) \equiv T$ gives rise to an L^2-symmetric

$$\mathbf{A}_\gamma = -J(\nabla_t - T\nabla R) \colon \Gamma(\gamma^*\xi) \to \Gamma(\gamma^*\xi),$$

where ∇ is any symmetric connection on M and $\mathbf{A}_\gamma$ does not depend on this choice. As proved in [HWZ95], the nontrivial eigenfunctions of $\mathbf{A}_\gamma$ have winding numbers (relative to τ) that depend only on their eigenvalues, defining a nondecreasing map from the spectrum $\sigma(\mathbf{A}_\gamma) \subset \mathbb{R}$ to $\mathbb{Z}$ that takes every value exactly twice (counting multiplicity of eigenvalues). One can therefore define the integers

$$\alpha_-^\tau(\gamma) = \max\left\{\mathrm{wind}^\tau(e) \,\middle|\, \mathbf{A}_\gamma e = \lambda e \text{ with } \lambda < 0\right\},$$
$$\alpha_+^\tau(\gamma) = \min\left\{\mathrm{wind}^\tau(e) \,\middle|\, \mathbf{A}_\gamma e = \lambda e \text{ with } \lambda > 0\right\},$$
$$p(\gamma) = \alpha_+^\tau(\gamma) - \alpha_-^\tau(\gamma).$$

Since γ is nondegenerate, 0 is not an eigenvalue of $\mathbf{A}_\gamma$; hence the **parity** $p(\gamma)$ is either 0 or 1, and [HWZ95] proves the relation

$$\mu_{CZ}^\tau(\gamma) = 2\alpha_-^\tau(\gamma) + p(\gamma) = 2\alpha_+^\tau(\gamma) - p(\gamma).$$

For this reason, the number $\alpha_-^\tau(\gamma)$ sometimes appears in the literature as $\lfloor \mu_{CZ}^\tau(\gamma)/2 \rfloor$.

Given a closed Reeb orbit γ, we denote its k-fold cover for $k \in \mathbb{N}$ by γ^k.

Remark C.1 The parity of Reeb orbits is closely related to the dichotomy between *elliptic* and *hyperbolic* orbits. Recall that since the linearized Reeb flow restricts to an ω-symplectic map on the transverse planes ξ along a periodic orbit γ, the product of the eigenvalues of this map is always 1. We call γ **elliptic** if the eigenvalues are a conjugate pair of nonreal numbers on the unit circle, and **hyperbolic** if they are both real but distinct from ± 1. (We exclude eigenvalues ± 1 from this dichotomy; in this case, either γ or γ^2 is degenerate.) If γ is an orbit whose covers are all nondegenerate, then one sees by taking

160   *A Quick Survey of Siefring's Intersection Theory*

powers of the eigenvalues that γ is elliptic if and only if all of its covers are elliptic. One can show, moreover, that γ has even parity if and only if both of the eigenvalues are positive; thus even orbits are always hyperbolic, and the same applies to all of their covers (see Exercise 3.18). It follows that elliptic orbits always have odd parity. Hyperbolic orbits with odd parity are sometimes also called **negative hyperbolic** orbits; their even covers have even parity and are referred to in the literature on symplectic field theory as *bad orbits*, for reasons having to do with orientations of moduli spaces (see, e.g., [Wenb, Chapter 11]).

We say that an almost complex structure J on $\mathbb{R} \times M$ is **compatible** with the stable Hamiltonian structure (ω, λ) if the following conditions hold:

- $J(\partial_r) = R$ for the coordinate vector field ∂_r in the $\mathbb{R}$-direction.
- $J(\xi) = \xi$ and $J|_\xi$ is compatible with $\omega|_\xi$.
- J is invariant under the translation action $(r, p) \mapsto (r + c, p)$ for all $c \in \mathbb{R}$.

More generally, we consider almost complex 4-manifolds $(\widehat{W}, J)$ with *cylindrical ends* as in [BEH$^+$03]. Concretely, this means $\widehat{W}$ decomposes into the union of a compact subset with a positive end $[0, \infty) \times M_+$ and a negative end $(-\infty, 0] \times M_-$, where $M_\pm$ are closed 3-manifolds equipped with stable Hamiltonian structures $(\omega_\pm, \lambda_\pm)$ and the restriction of J to each cylindrical end is compatible with these structures. This will be our standing assumption about $(\widehat{W}, J)$ in the following. For a punctured Riemann surface $(\dot{\Sigma}, j)$, we consider proper maps $u \colon \dot{\Sigma} \to \widehat{W}$ that are **asymptotically cylindrical** in the sense that they approximate trivial cylinders over closed Reeb orbits near each of their (positive or negative) nonremovable punctures; see §2.4 for a more precise definition of this term in the contact case.

C.2 The Intersection Pairing

Given the almost complex 4-manifold $(\widehat{W}, J)$ with cylindrical ends as described above, let τ denote a choice of trivialization for the complex line bundles $\xi_\pm = \ker \lambda_\pm$ along each simply covered closed Reeb orbit in $M_\pm$. This induces a trivialization of $\xi_\pm$ along *every* closed Reeb orbit by pulling back along multiple covers. The choice is arbitrary, but it is necessary in order to write down most formulas in the intersection theory, even though none of the important quantities depend on it. We assume $u \colon \dot{\Sigma} = \Sigma \backslash \Gamma_u \to \widehat{W}$ is a smooth asymptotically cylindrical map with positive and/or negative punctures $\Gamma_u = \Gamma_u^+ \cup \Gamma_u^- \subset \Sigma$, and for each puncture $z \in \Gamma_u$, let γ_z denote corresponding asymptotic Reeb

C.2 The Intersection Pairing 161

orbit. We also fix a second such map $v: \dot{\Sigma}' \to \widehat{W}$, denote its punctures by $\Gamma_v = \Gamma_v^+ \cup \Gamma_v^- \subset \Sigma'$ and use the same notation $\{\gamma_z\}_{z \in \Gamma_v}$ for its asymptotic orbits.[1]

Given any quantity $q_\pm(\gamma)$ that depends on both a Reeb orbit γ and a choice of sign $+$ or $-$, we will use the shorthand notation

$$\sum_{z \in \Gamma_u^\pm} q_\pm(\gamma_z) := \sum_{z \in \Gamma_u^+} q_+(\gamma_z) + \sum_{z \in \Gamma_u^-} q_-(\gamma_z).$$

A similar convention applies to summations over pairs of punctures in $\Gamma_u \times \Gamma_v$ with matching signs, and this will occur several times in the following.

The intersection product of two asymptotically cylindrical maps u and v is a symmetric pairing defined by

$$u * v := u \bullet_\tau v - \sum_{(z,\zeta) \in \Gamma_u^\pm \times \Gamma_v^\pm} \Omega_\pm^\tau(\gamma_z, \gamma_\zeta) \in \mathbb{Z}, \tag{C.1}$$

where the individual terms are defined as follows.

The **relative intersection number**

$$u \bullet_\tau v \in \mathbb{Z}$$

is the algebraic count of intersections between u and a generic perturbation of v that shifts it by an arbitrarily small positive distance in directions dictated by the chosen trivializations τ near infinity; hence the count is finite and depends only on the relative homology classes represnted by u and v and the homotopy class of the trivializations τ. The relative intersection number also appears in the ECH literature and is denoted there by $Q_\tau(u, v)$ (cf. [Hut02, Hut14]). Note that $u \bullet_\tau u$ is also well defined, and is sometimes denoted by $Q_\tau(u)$ in ECH.

The integers $\Omega_\pm^\tau(\gamma, \gamma')$ are defined for every pair of Reeb orbits γ, γ' and also depend on the trivializations τ. They satisfy $\Omega_\pm^\tau(\gamma, \gamma') = 0$ whenever γ and γ' are not covers of the same orbit, while for any simply covered orbit γ with integers $k, m \in \mathbb{N}$,

$$\Omega_\pm^\tau(\gamma^k, \gamma^m) := \min \left\{ \mp k\alpha_\mp^\tau(\gamma^m), \mp m\alpha_\mp^\tau(\gamma^k) \right\}. \tag{C.2}$$

The dependence on τ in the $\Omega_\pm^\tau$ terms cancels out the dependence in $u \bullet_\tau v$, so that $u * v$ is independent of τ; it is determined solely by the relative homology classes of u and v and their sets of asymptotic orbits. In particular, it is invariant under homotopies of u and v through families of smooth asymptotically cylindrical maps with fixed asymptotic orbits.

---

[1] Note that each of the orbits γ_z may be multiply covered, and the covering multiplicity is regarded as part of the data that defines γ_z.

162   *A Quick Survey of Siefring's Intersection Theory*

If u and v are also J-holomorphic and are not covers of the same simple curve, then we can also write

$$u * v = u \cdot v + \iota_\infty(u, v),$$

where both terms are nonnegative: the first denotes the actual algebraic count of intersections between u and v (of which the asymptotic results in [Sie08] imply there are only finitely many), and the second is an asymptotic contribution counting the number of "hidden" intersections that may emerge from infinity under a generic perturbation. A corollary is that if $u * v = 0$, then u and v are disjoint unless they cover the same simple curve. The converse of this is false in general, but one can use Fredholm theory with exponential weights to show that, for generic J, $\iota_\infty(u, v) = 0$ for all simple curves u and v belonging to some open and dense subsets of their respective moduli spaces.

To write down the asymptotic contribution $\iota_\infty(u, v)$ explicitly, one must first define its relative analogue $\iota_\infty^\tau(u, v)$, which depends only on the germ of u and v near infinity and on the trivializations τ. We have

$$\iota_\infty^\tau(u, v) = \sum_{(z,\zeta) \in \Gamma_u^\pm \times \Gamma_v^\pm} \iota_\infty^\tau(u, z; v, \zeta),$$

where for each pair of punctures $z \in \Gamma_u^\pm$ and $\zeta \in \Gamma_v^\pm$ with the same sign,

$$\iota_\infty^\tau(u, z; v, \zeta) \in \mathbb{Z}$$

is the algebraic count of intersections between $u|_{\mathcal{U}_z}$ and a generic perturbation of $v|_{\mathcal{U}_\zeta}$, with $\mathcal{U}_z$ and $\mathcal{U}_\zeta$ chosen to be suitably small neighborhoods of the respective punctures such that $u|_{\mathcal{U}_z}$ and $v|_{\mathcal{U}_\zeta}$ are disjoint, and the perturbation of $v|_{\mathcal{U}_\zeta}$ chosen to push it a small positive distance in directions dictated by the trivialization τ near infinity. The fact that this number is well defined depends on the existence of neighborhoods on which u and v are disjoint; hence it requires them to be geometrically distinct curves, and, of course, $\iota_\infty^\tau(u, z; v, \zeta) = 0$ whenever the asymptotic orbits γ_z and γ_ζ are disjoint. If, on the other hand, $\gamma_z = \gamma^k$ and $\gamma_\zeta = \gamma^m$ for some simply covered orbit γ and integers $k, m \in \mathbb{N}$, then $\iota_\infty^\tau(u, z; v, \zeta)$ can be computed in terms of the relative winding of v about u near infinity; a precise formula is derived in the discussion surrounding Equation (4.3). Combining this formula with the relative asymptotic analysis from [Sie08] then yields the bound $\iota_\infty^\tau(u, z; v, \zeta) \geq \Omega_\pm^\tau(\gamma_z, \gamma_\zeta)$, giving rise to the local asymptotic contribution

$$\iota_\infty(u, z; v, \zeta) := \iota_\infty^\tau(u, z; v, \zeta) - \Omega_\pm^\tau(\gamma_z, \gamma_\zeta),$$

which is independent of τ and is nonnegative, with equality if and only if all theoretical bounds on the winding of asymptotic eigenfunctions controlling the

C.3 The Adjunction Formula

approach of v to u at infinity are achieved. The geometric interpretation is that $\iota_\infty(u, z; v, \zeta)$ is the algebraic count of intersections between u and v that will appear in neighborhoods of these two punctures if u and v are perturbed to J'-holomorphic curves for some generic perturbation J' of J. The total number of hidden intersections is then

$$\iota_\infty(u, v) = \sum_{(z,\zeta) \in \Gamma_u^\pm \times \Gamma_v^\pm} \iota_\infty(u, z; v, \zeta).$$

C.3 The Adjunction Formula

The adjunction formula for a closed simple J-holomorphic curve $u: \Sigma \to W$ can be written as

$$[u] \cdot [u] = 2\delta(u) + c_N(u),$$

where $[u] \cdot [u] \in \mathbb{Z}$ denotes the homological self-intersection number of $[u] \in H_2(W)$, $c_N(u) := c_1([u]) - \chi(\Sigma)$ is the so-called *normal Chern number*, and $\delta(u)$ is the algebraic count of double points and critical points (cf. (2.3)). For a simple asymptotically cylindrical J-holomorphic curve $u: \dot{\Sigma} \to \widehat{W}$ with punctures Γ_u, the formula generalizes to

$$u * u = 2\left[\delta(u) + \delta_\infty(u)\right] + c_N(u) + \left[\bar{\sigma}(u) - \#\Gamma_u\right], \tag{C.3}$$

where $u * u$ is the intersection product defined in (C.1) with $u = v$, and the terms on the right-hand side will be explained in a moment. The most important thing to know about (C.3) is that the terms $u * u$, $c_N(u)$ and $\bar{\sigma}(u)$ are all homotopy invariant by definition, implying that $\delta(u) + \delta_\infty(u)$ is also homotopy invariant, while $\bar{\sigma}(u) - \#\Gamma_u$, $\delta(u)$ and $\delta_\infty(u)$ are always nonnegative. Moreover, as in the closed case, $\delta(u) = 0$ if and only if u is embedded. It follows that $\delta(u) + \delta_\infty(u) = 0$ gives a homotopy-invariant condition guaranteeing that u is embedded. The converse is false, as u can be embedded and have $\delta_\infty(u) > 0$, but one can again use Fredholm theory with exponential weights to show that generically the latter cannot happen for curves in some open and dense subset of the moduli space.

The **normal Chern number** is defined in the punctured case by

$$c_N(u) := c_1^\tau(u^* T\widehat{W}) - \chi(\dot{\Sigma}) + \sum_{z \in \Gamma_u^\pm} \pm \alpha_\mp^\tau(\gamma_z), \tag{C.4}$$

and it depends on the relative homology class of u and the topology of the domain $\dot{\Sigma}$, but not on the trivializations τ. Here $c_1^\tau(u^* T\widehat{W})$ denotes the **relative first Chern number** of the complex vector bundle $u^* T\widehat{W} \to \dot{\Sigma}$ with respect

164   *A Quick Survey of Siefring's Intersection Theory*

to the natural trivializations at infinity induced by τ. Recall that if $E \to \dot{\Sigma}$ is a complex line bundle equipped with a preferred trivialization τ_E near infinity, one can define $c_1^{\tau_E}(E) \in \mathbb{Z}$ as the algebraic count of zeroes of any generic section of E that is constant and nonzero with respect to τ_E near infinity. The relative first Chern number of higher-rank bundles is then defined via the direct sum property $c_1^{\tau_E \oplus \tau_F}(E \oplus F) = c_1^{\tau_E}(E) + c_1^{\tau_F}(F)$. Since $u^*T\widehat{W}$ has a natural splitting over the positive/negative cylindrical ends into the direct sum of a trivial complex line bundle with $\xi_\pm = \ker \lambda_\pm$, τ naturally induces a trivialization of $u^*T\widehat{W}$ over the ends, and we define $c_1^\tau(u^*T\widehat{W})$ accordingly. (The same quantity is often denoted by $c_\tau(u)$ in the ECH literature; see [Hut02, Hut14].) The normal Chern number is often most convenient to calculate via the formula

$$2c_N(u) = \text{ind}(u) - 2 + 2g + \#\Gamma_{\text{even}}, \tag{C.5}$$

where $\text{ind}(u)$ denotes the virtual dimension of the moduli space containing u (see (A.5)), g is the genus of its domain, and $\Gamma_{\text{even}} \subset \Gamma_u$ is the set of punctures $z \in \Gamma_u$ that satisfy $p(\gamma_z) = 0$; i.e., the Conley–Zehnder index of the corresponding Reeb orbit is even. This relation is an easy consequence of the Fredholm index formula and the usual relations between Conley–Zehnder indices and the winding numbers $\alpha_\pm^\tau(\gamma)$; cf. (3.18). The proper interpretation of $c_N(u)$ is as a homotopy-invariant algebraic count of zeroes of the normal bundle of an immersed perturbation of u, including zeroes that are "hidden at infinity" but may emerge under small perturbations of u as a holomorphic curve.

The term $\bar{\sigma}(u)$ is called the **spectral covering number** and is a sum of terms

$$\bar{\sigma}(u) := \sum_{z \in \Gamma_u^\pm} \bar{\sigma}_\mp(\gamma_z),$$

each of which is a positive integer that depends only on the orbit γ_z and can be greater than 1 only if γ_z is multiply covered. Specifically, for any simply covered orbit γ and $k \in \mathbb{N}$, $\bar{\sigma}_\pm(\gamma^k)$ is the covering multiplicity of any of the nontrivial asymptotic eigenfunctions e of $\mathbf{A}_{\gamma^k}$ that satisfy $\text{wind}^\tau(e) = \alpha_\pm^\tau(\gamma^k)$. It turns out that the dependence of $\bar{\sigma}_\pm(\gamma^k)$ on the orbit γ is fairly mild, as one can show that

$$\bar{\sigma}_\pm(\gamma^k) = \gcd(k, \alpha_\pm^\tau(\gamma^k)) \tag{C.6}$$

(cf. Remark 4.3.) Thus $\bar{\sigma}(u) - \#\Gamma$ vanishes, for instance, whenever all the asymptotic orbits of u are simply covered.

The **singularity index** $\delta(u)$ is defined just as in the closed case, as a signed count of double points of u plus positive contributions for each critical point, interpreted as the count of double points that appear near each critical point after an immersed perturbation (cf. Lemma 2.6). The only difference from the closed case is that since $\dot{\Sigma}$ is noncompact, it is less obvious that $\delta(u)$ is well defined, but the relative asymptotic results of [Sie08] imply that double points and critical points of a simple curve cannot occur near infinity; hence $\delta(u)$ is finite.

The term $\delta_\infty(u)$ is an algebraic count of "hidden" double points; i.e., it is the number of extra contributions to $\delta(u)$ that will emerge from infinity if u is perturbed to a J'-holomorphic curve for a generic perturbation J' of J. There are two possible sources of such hidden double points: first, any pair of distinct punctures $z, \zeta \in \Gamma_u^\pm$ with the same sign such that the corresponding asymptotic orbits γ_z and γ_ζ are identical up to multiplicity contributes $\iota_\infty(u, z; u, \zeta)$ as in the definition of $u * v$. Note that $\iota_\infty(u, z; u, \zeta)$ is well defined as long as u is simple and $z \neq \zeta$, since the two punctures then have neighborhoods $\mathcal{U}_z$ and $\mathcal{U}_\zeta$ such that $u(\mathcal{U}_z) \cap u(\mathcal{U}_\zeta) = \varnothing$. Additional hidden intersections can emerge from any single puncture z such that γ_z is multiply covered, since u in the neighborhood of such a puncture has multiple branches that become arbitrarily close to each other near infinity. Denoting the contribution from such punctures by $\delta_\infty(u, z)$, we have

$$\delta_\infty(u) = \frac{1}{2} \sum_{z,\zeta \in \Gamma_u^\pm, \ z \neq \zeta} \iota_\infty(u, z; u, \zeta) + \sum_{z \in \Gamma_u^\pm} \delta_\infty(u, z).$$

In particular, $\delta_\infty(u) = 0$ whenever all asymptotic orbits of u are distinct and simply covered, though it can also be zero without this condition. As with $\iota_\infty(u, z; v, \zeta)$, writing down a precise formula for $\delta_\infty(u, z)$ requires first defining a relative version that depends on the trivialization τ: we define

$$\iota_\infty^\tau(u, z) \in \mathbb{Z}$$

as the algebraic count of intersections between $u|_{\mathcal{U}_z}$ and a generic small perturbation of itself, where $\mathcal{U}_z$ is a neighborhood of z on which u is embedded and the perturbation is chosen to shift u a small positive distance in directions dictated by τ. As with $\iota_\infty^\tau(u, z; v, \zeta)$, one can compute $\iota_\infty^\tau(u, z)$ in terms of the winding numbers of asymptotic eigenfunctions that control the relative approach of different branches of $u|_{\mathcal{U}_z}$ to each other near infinity (cf. (4.6)). One derives from this the theoretical bound $\iota_\infty^\tau(u, z) \geq \Omega_\pm^\tau(\gamma_z)$, where for any simply covered orbit γ and $k \in \mathbb{N}$,

$$\Omega_\pm^\tau(\gamma^k) := \mp(k - 1)\alpha_\mp^\tau(\gamma^k) + \left[\bar{\sigma}_\mp(\gamma^k) - 1\right]. \tag{C.7}$$

The precise definition of $\delta_\infty(u, z)$ is then

$$\delta_\infty(u, z) := \frac{1}{2} \left[\iota_\infty^\tau(u, z) - \Omega_\pm^\tau(\gamma_z)\right],$$

which is a nonnegative integer and is independent of τ.

166 *A Quick Survey of Siefring's Intersection Theory*

As mentioned in Remark 4.14, the computation of $\iota_\infty^\tau(u, z)$ in terms of winding numbers also leads to an alternative interpretation of it as the writhe of a braid, which we will say more about in §C.6. Up to issues of bookkeeping, (C.3) is also equivalent to the so-called *relative* adjunction formula first written down by Hutchings, see (in particular, [Hut02, Remark 3.2]). The innovation of [Sie11] was to transform this into a relation between homotopy-invariant quantities that have geometric meanings independent of any choice of trivializations.

C.4 Covering Relations

We now state a few useful results about multiply covered holomorphic curves that are not mentioned elsewhere in this book, but are easy to prove based on the definitions given above. The results of the present section are due to the author, and complete proofs may be found in [Wen10a, §4.2].

If u and v are two closed J-holomorphic curves in a closed almost complex 4-manifold and $\widetilde{u}$ is a d-fold multiple cover of u, then the relation

$$[\widetilde{u}] \cdot [v] = d[u] \cdot [v]$$

is obvious since $[\widetilde{u}] = d[u] \in H_2(W)$. Things are less straightforward in the punctured case because $u * v$ depends on more than just homology, and, e.g., Exercise 4.19 exhibits a specific scenario in which the $*$-product fails to satisfy the obvious analogue of the above relation. One can still, however, prove the following:

Proposition C.2 *Suppose u, $\widetilde{u}$ and v are asymptotically cylindrical J-holomorphic curves in $\widehat{W}$ such that $\widetilde{u}$ is a d-fold cover of u for $d \in \mathbb{N}$. Then*

$$\widetilde{u} * v \geq d(u * v).$$

The proof of this inequality is based on the formula (C.1), in which the relative intersection numbers are easily seen to satisfy the straightforward relation $\widetilde{u} \bullet_\tau v = d(u \bullet_\tau v)$, and thus the tricky part is to understand what happens to the terms $\Omega_\pm^\tau(\gamma_z, \gamma_\zeta)$ when each of the orbits γ_z is replaced by a collection of covers of γ_z whose multiplicities add up to d. The answer is a bit intricate if one aims to write it down precisely, because the winding numbers $\alpha_\pm^\tau(\gamma)$ do not in general behave linearly with respect to iteration of the orbit, but for the purposes of the inequality in Proposition C.2, the information in the following lemma suffices. This lemma is closely related to Exercise 3.18, and it can be derived

C.4 Covering Relations

from the properties of asymptotic eigenfunctions and their winding numbers proved in [HWZ95] (in particular Theorem 3.15).

Lemma C.3 *For every closed Reeb orbit γ and every $k \in \mathbb{N}$, there exist integers $q_{\pm}(\gamma; k) \in \{0, \ldots, k - 1\}$ such that*

$$\alpha_{\pm}(\gamma^k) = k\alpha_{\pm}(\gamma) \mp q_{\pm}(\gamma; k).$$

It is also sometimes useful to have a similar covering relation for the normal Chern number, since the latter appears in the adjunction formula. Recall that, if $\varphi \colon (\Sigma', j') \to (\Sigma, j)$ is a d-fold holomorphic branched cover, then the Riemann–Hurwitz formula gives

$$- \chi(\Sigma') + d\chi(\Sigma) = Z(d\varphi), \tag{C.8}$$

where $Z(d\varphi)$ denotes the algebraic count of branch points of φ,

$$Z(d\varphi) := \sum_{z \in d\varphi^{-1}(0)} \mathrm{ord}(d\varphi; z) \geq 0.$$

One easy proof of this formula views $d\varphi$ as a section of the line bundle $\mathrm{Hom}_{\mathbb{C}}(T\Sigma', \varphi^*T\Sigma)$, whose first Chern number is the left-hand side of (C.8). If $u \colon (\Sigma, j) \to (W, J)$ is a closed J-holomorphic curve and $\widetilde{u} = u \circ \varphi$, this leads to the relation

$$c_N(\widetilde{u}) = d \cdot c_N(u) + Z(d\varphi).$$

In the punctured case, one can easily show that (C.8) continues to hold for a branched cover of punctured surfaces, but additional terms appear in the normal Chern number due to the fact that $\alpha_{\pm}^{\tau}(\gamma^d) \neq d\alpha_{\pm}^{\tau}(\gamma)$ in general. As in Proposition C.2, the result is then most easily stated as an inequality.

Proposition C.4 *Suppose u and $\widetilde{u} = u \circ \varphi$ are asymptotically cylindrical J-holomorphic curves in $\widehat{W}$, where $\varphi \colon \dot{\Sigma}' \to \dot{\Sigma}$ is a d-fold holomorphic branched cover of punctured Riemann surfaces whose algebraic count of branch points is $Z(d\varphi) \geq 0$. Then*

$$c_N(\widetilde{u}) \geq d \cdot c_N(u) + Z(d\varphi).$$

Remark C.5 One can extract from the proofs in [Wen10a, §4.2] various conditions to characterize when the inequalities in Propositions C.2 and C.4 are strict or not. The easiest comes from the observation that the integers $q_{\pm}(\gamma; k)$ in Lemma C.3 vanish whenever γ has even parity. It follows, for instance, that Proposition C.2 becomes an equality whenever every simple Reeb orbit that has a cover appearing among the asymptotic orbits of u and v (with the same sign!) is even. Similarly, Proposition C.4 is an equality if all the asymptotic orbits of u are even.

168   *A Quick Survey of Siefring's Intersection Theory*

C.5 The Intersection Product of Buildings

Another topic not mentioned elsewhere in this book is the extension of the $*$-pairing to the *compactified* moduli space $\overline{\mathcal{M}}_g(\widehat{W}, J)$ of holomorphic buildings defined in [BEH$^+$03]. Following [Sie11], one can define this in fairly general terms as follows.

If $u \in \overline{\mathcal{M}}_g(\widehat{W}, J)$ and $v \in \overline{\mathcal{M}}_{g'}(\widehat{W}, J)$ are two nodal J-holomorphic curves in $\widehat{W}$, i.e., holomorphic buildings with no upper or lower levels, then the definition of $u * v$ requires no change from before. Recall that the domain of a punctured nodal curve u is a possibly disconnected punctured Riemann surface $\dot{S}$ endowed with a finite set of points $\Delta \subset \dot{S}$, the **nodal points**, which are grouped into pairs on which u has matching values (see Appendix A.1). A nodal curve then belongs to $\overline{\mathcal{M}}_g(\widehat{W}, J)$ if the surface obtained by performing connected sums on $\dot{S}$ at each of the nodal pairs is connected with genus g and every component of $\dot{S} \backslash \Delta$ on which u is constant has negative Euler characteristic. For the present discussion, there is no need to impose either of these conditions, and thus we are free to consider nodal curves that are nonstable and/or disconnected (even after gluing together their nodes). If $u \colon \dot{S} \to \widehat{W}$ is a nodal curve and $\dot{S}_0 \subset \dot{S}$ is a connected component of its domain (ignoring nodes), then let us call the restriction $u|_{\dot{S}_0} \colon \dot{S}_0 \to \widehat{W}$ a **connected component** of u. Now it is easy to check that if u and v are nodal curves whose connected components are $u_1, \ldots, u_m$ and $v_1, \ldots, v_n$, respectively, the $*$-pairing is additive in the obvious way, namely

$$u * v = \sum_{i=1}^m \sum_{j=1}^n u_i * v_j.$$

Things become more interesting if we consider buildings with multiple levels. Suppose $(\widehat{W}_0, J_0)$ and $(\widehat{W}_1, J_1)$ are two almost complex 4-manifolds with cylindrical ends such that the positive end of $(\widehat{W}_0, J_0)$ matches the negative end of $(\widehat{W}_1, J_1)$, meaning that the underlying 3-manifolds and stable Hamiltonian structures are the same, and so are the restrictions of J_0 and J_1 to translation-invariant almost complex structures on the relevant ends. We will then refer to the symbol $\widehat{W}_0 \odot \widehat{W}_1$ as the **concatenation** of $\widehat{W}_0$ with $\widehat{W}_1$, and say that $u_0 \odot u_1$ is a **holomorphic building** in $\widehat{W}_0 \odot \widehat{W}_1$ if u_0 and u_1 are (possibly disconnected and/or nodal) asymptotically cylindrical holomorphic curves in $(\widehat{W}_0, J_0)$ and $(\widehat{W}_1, J_1)$, respectively, equipped with the extra structure of a bijection between the positive punctures of u_0 and the negative punctures of u_1 that sends each puncture to one that has the same asymptotic orbit. We shall refer to u_0 and u_1 as the (lower and upper) **levels** of $u_0 \odot u_1$ and call the Reeb orbits along which they connect to each other **breaking orbits**. These definitions extend in an

C.5 The Intersection Product of Buildings 169

obvious way to allow concatenations with more than two inputs, making $\odot$ an associative operation. In this language, $\overline{\mathcal{M}}_g(\widehat{W}, J)$ consists of all holomorphic buildings in

$$(\mathbb{R} \times M_-) \odot \cdots \odot (\mathbb{R} \times M_-) \odot \widehat{W} \odot (\mathbb{R} \times M_+) \odot \cdots \odot (\mathbb{R} \times M_+)$$

that are connected with arithmetic genus g and satisfy the usual stability condition, where $\mathbb{R} \times M_\pm$ is an abbreviation for the symplectization of $M_\pm$ with the same $\mathbb{R}$-invariant almost complex structure that appears at the corresponding end of $\widehat{W}$, and any nonnegative numbers of such symplectization levels are allowed to appear in the concatenation.

Recall that the stability condition on elements of $\overline{\mathcal{M}}_g(\widehat{W}, J)$ precludes (among other things) the existence of any level that lives in an $\mathbb{R}$-invariant symplectization and consists of nothing more than a disjoint union of orbit cylinders with no nodes. It is necessary to exclude buildings that don't satisfy this condition in order for the natural topology on $\overline{\mathcal{M}}_g(\widehat{W}, J)$ to be Hausdorff, but, for our present purposes, it will be useful to avoid imposing any such requirement on buildings in concatenations. We are then free to define the following operation: given a building $u = u_1 \odot \cdots \odot u_N$ in $\widehat{W}_1 \odot \cdots \odot \widehat{W}_N$ and $k \in \{0, \ldots, N\}$, we construct the building

$$u' = u_1 \odot \cdots \odot u_k \odot v \odot u_{k+1} \odot \cdots \odot u_N \quad \text{in} \quad \widehat{W}_1 \odot \cdots \odot \widehat{W}_k \odot \widehat{V} \odot \widehat{W}_{k+1} \odot \cdots \odot \widehat{W}_N,$$

where $\widehat{V}$ is the symplectization corresponding to the positive end of $\widehat{W}_k$ and negative end of $\widehat{W}_{k+1}$ and v is a disjoint union of orbit cylinders in $\widehat{V}$, one for each of the breaking orbits that connect u_k to u_{k+1}. Here the cases $k = 0$ and $k = N$ are also allowed in order to accommodate adding a trivial level at the very bottom or top of the building, and one should also keep in mind that v could be an *empty* curve – this is the case if u_k has no positive ends and u_{k+1} has no negative ends. Any building obtained from u by a finite sequence of such operations will be called an **extension** of u.

Given two buildings $u = u_1 \odot \cdots \odot u_N$ and $v = v_1 \odot \cdots \odot v_N$ in a concatenation $\widehat{W}_1 \odot \cdots \odot \widehat{W}_N$, we make an arbitrary choice of trivializations τ along all closed Reeb orbits at the ends of each of $\widehat{W}_1, \ldots, \widehat{W}_N$ and define

$$u * v := \sum_{i=1}^{N} u_i \bullet_\tau v_i - \sum_{(z,\zeta) \in \Gamma_{u_N}^+ \times \Gamma_{v_N}^+} \Omega_+^\tau(\gamma_z, \gamma_\zeta) - \sum_{(z,\zeta) \in \Gamma_{u_1}^- \times \Gamma_{v_1}^-} \Omega_-^\tau(\gamma_z, \gamma_\zeta). \quad \text{(C.9)}$$

Here the dependence on τ at the positive ends of the top level and negative ends of the bottom level is canceled by the $\Omega_\pm^\tau$ terms for the same reasons as in (C.1), while changing τ at any of the breaking orbits between levels $k - 1$ and k alters $u_{k-1} \bullet_\tau v_{k-1}$ and $u_k \bullet_\tau v_k$ in ways that cancel out; thus the total expression

170   *A Quick Survey of Siefring's Intersection Theory*

is independent of choices. This definition does not yet allow us to define $u * v$ for an arbitrary pair of stable buildings $u \in \overline{\mathcal{M}}_g(\widehat{W}, J)$ and $v \in \overline{\mathcal{M}}_{g'}(\widehat{W}, J)$, because these may, in general, have differing numbers of levels. However, one can always add extra trivial symplectization levels to one or both of them to produce a pair of *nonstable* buildings that live in the same concatenation of cobordisms. With this understood, we define

$$u * v := u' * v' \in \mathbb{Z} \quad \text{for} \quad u \in \overline{\mathcal{M}}_g(\widehat{W}, J) \text{ and } v \in \overline{\mathcal{M}}_{g'}(\widehat{W}, J), \qquad \text{(C.10)}$$

where u' and v' are any choices of extensions of u and v that make $u' * v'$ well defined in the sense of (C.9).

Proposition C.6   *The pairing $u * v$ defined in (C.10) for stable holomorphic buildings in $\widehat{W}$ has the following properties:*

(1) *It is independent of the choices of extensions u' and v'.*
(2) *It is continuous with respect to the natural topologies on $\overline{\mathcal{M}}_g(\widehat{W}, J)$ and $\overline{\mathcal{M}}_{g'}(\widehat{W}, J)$; e.g., if $u_k \in \mathcal{M}_g(\widehat{W}, J)$ is a sequence converging to a building $u \in \overline{\mathcal{M}}_g(\widehat{W}, J)$ in the sense of [BEH$^+$03], then $u_k * v = u * v$ for large k.*
(3) *It is superadditive with respect to concatenation; i.e., for any (not necessarily stable) buildings u_-, v_- and u_+, v_+ such that the concatenations $u_- \odot u_+$ and $v_- \odot v_+$ and the intersection numbers $u_\pm * v_\pm$ in the sense of (C.9) are well defined, one has*

$$(u_- \odot u_+) * (v_- \odot v_+) \geq u_- * v_- + u_+ * v_+,$$

with equality whenever all the simple orbits with covers appearing as breaking orbits in both $u_- \odot u_+$ and $v_- \odot v_+$ are even, and strict inequality if any of these simple orbits are elliptic.

Remark C.7   We have stated the above result with reference to one of the three compactified moduli spaces of holomorphic buildings defined in [BEH$^+$03]; i.e., for the degeneration of curves in a completed *nontrivial* symplectic cobordism. The result can be adapted in obvious ways for the other two scenarios, namely for degenerations of curves in a symplectization (so that each level is defined only up to $\mathbb{R}$-translation and there is no distinguished "main level"), and degenerations with respect to neck-stretching. In the symplectization case, the freedom to choose different extensions of u and v is more useful than one might at first imagine.

For example, in Lemma 5.15 we considered two curves u_P and v in a symplectization $\mathbb{R} \times M$, where u_P was a page of a holomorphic open book (which has only positive punctures) and v was any other curve whose positive ends are all asymptotic to simple orbits in the binding of the open book. Having shown in the previous lemma that $u_P * u_\gamma = 0$ for all of the trivial cylinders

C.5 The Intersection Product of Buildings

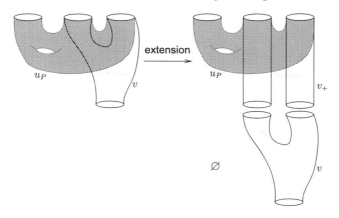

Figure C.1 An alternative to the homotopy argument depicted in Figure 5.4 of Lecture 5, using the intersection number between buildings to prove Lemma 5.15.

u_γ over asymptotic orbits γ of u_P, we then used a homotopy of asymptotically cylindrical maps (Figure 5.4 in Lecture 5) to prove that $u * v$ is a sum of such terms, and therefore vanishes. An alternative argument for the second step is illustrated in Figure C.1: define extensions u'_P of u_P and v' of v as buildings in $(\mathbb{R} \times M) \odot (\mathbb{R} \times M)$, where $u'_P := \emptyset \odot u_P$ has a trivial level added below the original curve (the trivial level is the empty curve since u_P has no negative ends), and $v' := v \odot v_+$ with v_+ as the disjoint union of trivial cylinders over the positive asymptotic orbits of v. Instead of exploiting homotopy invariance as we did in Lemma 5.15, one could now apply Proposition C.6 and write

$$u_P * v = u'_P * v' \geq (\emptyset * v) + (u_P * v_+) = u_P * v_+,$$

where the last equality follows since the empty curve has zero intersection number with everything else. The inequality is in this case an equality because every simple orbit that has a cover appearing as a breaking orbit of both u'_P and v' is even – this is a statement about the empty set, and is therefore true. This proves $u_P * v = u_P * v_+$, and the latter again vanishes due to Lemma 5.14. Morally, one can think of the replacement of v with v_+ as an "unbounded homotopy"; i.e., it shifts v by $\mathbb{R}$-translation infinitely far downward so that v now occupies a lower level. In this sense, the independence of $u * v$ with respect to choices of extensions is just another manifestation of homotopy invariance.

The possibility of strict inequality in the third item of Proposition C.6 reveals another interesting "hidden intersection" phenomenon: intersections between buildings can be hidden in the breaking orbits between levels. Concretely, suppose u_k and v_k are two sequences of smooth curves in the completed cobordism

172   *A Quick Survey of Siefring's Intersection Theory*

$\widehat{W}$ that converge to two-level buildings $u = u_- \odot u_+$ and $v = v_- \odot v_+$, respectively, in $\widehat{W} \odot (\mathbb{R} \times M_+)$ such that $u_\pm$ and $v_\pm$ are disjoint and satisfy $u_\pm * v_\pm = 0$. Then the curves in each individual level do not have any hidden intersections, meaning one could make arbitrary small perturbations of the data on $\widehat{W}$ or $\mathbb{R} \times M_+$ and rely on $u_\pm$ remaining disjoint from $v_\pm$. But it is nonetheless possible that u_k and v_k intersect each other for all k large, in which case these intersections must escape from every compact subset of $\widehat{W}$ as $k \to \infty$, so as not to survive as intersections of u_- with v_-. At the same time, the intersections of u_k with v_k cannot congregate as $k \to \infty$ in any compact subset of $\mathbb{R} \times M_+$ after shifting the positive cylindrical end to focus on the convergence of the upper level, as otherwise they would survive as intersections of u_+ with v_+. Instead, intersections congregate in the increasingly wide area "between levels" as $k \to \infty$, so that they do not appear at all in the limit. Despite this, they are accounted for by $u * v$, which must in this case be strictly larger than $u_- * v_- + u_+ * v_+ = 0$. This phenomenon has sometimes been exploited in applications, e.g., to define a version of contact homology on the complement of a set of fixed Reeb orbits (see [HMS15]). Relatedly, one can use a "local" version of the adjunction formula to show that the breaking orbits of a single building with embedded levels can also hide double points of simple curves that degenerate to them (see [CW]).

Looking at the definition (C.9), one sees that in the context of Proposition C.6,

$$(u_- \odot u_+) * (v_- \odot v_+) - u_- * v_- - u_+ * v_+ = \sum_{(z,\zeta)} \mathrm{Br}(\gamma_z, \gamma_\zeta), \qquad (C.11)$$

where

$$\mathrm{Br}(\gamma, \gamma') := \Omega^\tau_+(\gamma, \gamma') + \Omega^\tau_-(\gamma, \gamma'),$$

and the sum is over all pairs of breaking punctures, i.e., all (z, ζ) with $z \in \Gamma^+_{u_-} \cong \Gamma^-_{u_+}$ and $\zeta \in \Gamma^+_{v_-} \cong \Gamma^-_{v_+}$. The so-called **breaking contribution** $\mathrm{Br}(\gamma_z, \gamma_\zeta) \in \mathbb{Z}$ is independent of the choice of trivialization τ and vanishes if γ_z and γ_ζ are disjoint, whereas if $\gamma_z = \gamma^k$ and $\gamma_\zeta = \gamma^m$ for some $k, m \in \mathbb{N}$ and a simple orbit γ, one extracts from (C.2) the formula

$$\mathrm{Br}(\gamma^k, \gamma^m) = \min\left\{k\alpha^\tau_+(\gamma^m), m\alpha^\tau_+(\gamma^k)\right\} - \max\left\{k\alpha^\tau_-(\gamma^m), m\alpha^\tau_-(\gamma^k)\right\}.$$

Using the relation $p(\gamma) = \alpha^\tau_+(\gamma) - \alpha^\tau_-(\gamma)$, it is now a straightforward exercise to prove the inequality

$$\min\left\{kp(\gamma^m), mp(\gamma^k)\right\} \le \mathrm{Br}(\gamma^k, \gamma^m) \le \max\left\{kp(\gamma^m), mp(\gamma^k)\right\}. \qquad (C.12)$$

The breaking contributions are thus manifestly nonnegative; moreover, they vanish whenever γ is even and are strictly positive if γ is elliptic and all its

C.5 The Intersection Product of Buildings 173

covers are nondegenerate, since the latter guarantees that the covers are also odd (see Remark C.1). This is the reason for the inequality stated in Proposition C.6.

Remark C.8 If u_γ and $u_{\gamma'}$ denote the trivial cylinders over two Reeb orbits γ and γ', then taking the same set of trivializations at positive and negative ends always gives $u_\gamma \bullet_\tau u_{\gamma'} = 0$ since $u_{\gamma'}$ admits a global perturbation that is compatible with τ near infinity and everywhere disjoint from u_γ. Plugging in the definition of the $*$-pairing, one obtains from this a geometric interpretation of the breaking contribution $\mathrm{Br}(\gamma, \gamma')$, namely

$$\mathrm{Br}(\gamma, \gamma') = -u_\gamma * u_{\gamma'}, \tag{C.13}$$

along with the useful corollary that $u_\gamma * u_{\gamma'}$ is never positive (cf. Exercise 4.19).

An analogue of Proposition C.6 for the normal Chern number is sometimes needed for applications of the adjunction formula. For a nodal curve $u \colon \dot{S} \to \hat{W}$ with nodal points $\Delta \subset \dot{S}$, (C.4) is not quite the right definition because $\chi(\dot{S})$ does not generally match the Euler characteristic of the surface obtained from $\dot{S}$ by performing connected sums at all nodal pairs. To achieve this and thus ensure that c_N is continuous under degenerations from smooth curves to nodal curves, one defines

$$c_N(u) := c_1^\tau(u^* T\hat{W}) - \chi(\dot{S} \backslash \Delta) + \sum_{z \in \Gamma_u^+} \alpha_-^\tau(\gamma_z) - \sum_{z \in \Gamma_u^-} \alpha_+^\tau(\gamma_z). \tag{C.14}$$

If u has connected components $u_1, \ldots, u_m$, we then have the relation

$$c_N(u) = \sum_{i=1}^m c_N(u_i) + 2\,(\#\Delta). \tag{C.15}$$

For a building $u_1 \odot \cdots \odot u_N$ in $\hat{W}_1 \odot \cdots \odot \hat{W}_N$, the above definition now generalizes naturally as

$$c_N(u_1 \odot \cdots \odot u_N) := \sum_{k=1}^N \left[c_1^\tau(u_k^* T\hat{W}_k) - \chi(\dot{\Sigma}_k \backslash \Delta_k) \right]$$
$$+ \sum_{z \in \Gamma_{u_N}^+} \alpha_-^\tau(\gamma_z) - \sum_{z \in \Gamma_{u_1}^-} \alpha_+^\tau(\gamma_z), \tag{C.16}$$

where for each $k = 1, \ldots, N$, $\dot{\Sigma}_k$ denotes the (possibly disconnected) domain of the level u_k, and $\Delta_k \subset \dot{\Sigma}$ is the set of nodal points in that domain. This leads to the following proposition:

Proposition C.9 *The normal Chern number is continuous with respect to the natural topology of $\overline{\mathcal{M}_g}(\hat{W}, J)$. Moreover, it is superadditive with respect to*

174   *A Quick Survey of Siefring's Intersection Theory*

concatenation: in particular, for any pair of (not necessarily stable) buildings u_- *and* u_+ *for which the concatenation* $u_- \odot u_+$ *is well defined, one has*

$$c_N(u_- \odot u_+) = c_N(u_-) + c_N(u_+) + \sum_z p(\gamma_z) \geq c_N(u_-) + c_N(u_+),$$

where the sum is over all breaking punctures that connect u_- *to* u_+, *i.e.,* $z \in \Gamma_{u_-}^+ \cong \Gamma_{u_+}^-$.

C.6 Comparison with the ECH Literature

Intersection theory plays a major role in Hutchings's theory of embedded contact homology (ECH), and, in fact, early developments in ECH (notably the paper [Hut02]) provided some of the inspiration behind Siefring's intersection theory of punctured holomorphic curves. Though the $*$-pairing and Siefring's adjunction formula do not usually appear in papers on ECH, the relative intersection numbers and relative adjunction formula appear quite prominently, with differing notational conventions, and many of the same winding bounds that underlie Siefring's theory also play crucial roles in ECH. The aim of this section is to provide a glossary for translating between these two contexts.

Aside from notation, the major difference between the ECH literature and our treatment in this book is that ECH expresses all relative asymptotic quantities such as $\iota_\infty^\tau(u, v)$ and $\iota_\infty^\tau(u)$ in terms of topological invariants of certain braids. Concretely, for two asymptotically cylindrical curves u and v that do not have identical images, we have written the count of intersections near infinity that appear under small perturbations moving v in the direction of asymptotic trivializations τ as

$$\iota_\infty^\tau(u, v) = \sum_{(z,\zeta) \in \Gamma_u^\pm \times \Gamma_v^\pm} \iota_\infty^\tau(u, z\,; v, \zeta),$$

where $\iota_\infty^\tau(u, z\,; v, \zeta) \in \mathbb{Z}$ denotes the contribution coming from the specific punctures $z \in \Gamma_u^\pm$ and $\zeta \in \Gamma_v^\pm$. The latter can only be nonzero if γ_z and γ_ζ are covers of the same underlying simple orbit γ; thus let us assume this henceforth. We wrote down Siefring's formula for $\iota_\infty^\tau(u, z\,; v, \zeta)$ in terms of relative winding numbers in (4.3). Hutchings expresses the same formula as follows: Writing u and v in holomorphic cylindrical coordinates $(s, t) \in Z_\pm$ near the punctures z and ζ, respectively,[2] we can fix some $s_0 \gg 0$ and consider the

---

[2] Recall that we denote $Z_+ := [0, \infty) \times S^1$ and $Z_- := (-\infty, 0] \times S^1$, where the convention is to use the former near positive punctures and the latter near negative punctures.

C.6 Comparison with the ECH Literature 175

restrictions of u and v to $\{\pm s_0\} \times S^1 \subset Z_\pm$; this defines a disjoint pair of (possibly multiply covered) oriented loops in the positive or negative cylindrical end of $\widehat{W}$. Projecting them to the 3-manifold $M_\pm$ then gives disjoint oriented loops $\beta_z, \beta_\zeta : S^1 \to M_\pm$ that live in an arbitrarily small tubular neighborhood of γ. If we now use the trivialization τ to identify the neighorhood of γ with $S^1 \times \mathbb{D}$, then β_z and β_ζ become a pair of disjoint braids – strictly speaking, they are in general "multiply covered braids," but one can perturb to make each of them embedded and thus view them as honest braids without changing any essential features of this discussion. The **linking number** between these two braids,

$$\ell_\tau(\beta_z, \beta_\zeta) \in \mathbb{Z},$$

is defined as one-half the signed number of crossings of strands of β_z with strands of β_ζ, where the sign convention is that counterclockwise twists count positively. (As mentioned in [Hut02, §3.1], this convention differs from much of the knot theory literature, but it is used consistently in papers on ECH and we shall stick with it here as well.) The precise relation between this linking number and Siefring's relative asymptotic intersection numbers is then given by

$$\iota^\tau_\infty(u, z; v, \zeta) = \mp\ell_\tau(\beta_z, \beta_\zeta), \tag{C.17}$$

where the sign $\mp$ is opposite the signs $\pm$ of the two punctures.

The relative adjunction formula in (4.7) also includes the term

$$\iota^\tau_\infty(u) = \sum_{z, \zeta \in \Gamma^\pm, z \neq \zeta} \iota^\tau_\infty(u, z; u, \zeta) + \sum_{z \in \Gamma^\pm} \iota^\tau_\infty(u, z),$$

which is defined only when u is a simple curve; the contribution $\iota^\tau_\infty(u, z) \in \mathbb{Z}$ for each puncture z can only be nonzero when the orbit γ_z is multiply covered, as it is the count of intersections in a neighborhood of z between u and a small perturbation of itself that is pushed in the direction of the trivialization τ near infinity. Siefring's formula for $\iota^\tau_\infty(u, z)$ in terms of winding numbers appears in (4.6), and its topological interpretation is (up to a sign) as the **writhe** of the braid β_z described in the previous paragraph,

$$w_\tau(\beta_z) \in \mathbb{Z}.$$

Here the fact that u is simple guarantees that it is embedded in a neighborhood of the puncture z; thus the braid β_z is automatically embedded, and its writhe is defined as the signed number of crossings of strands, using the same sign convention mentioned above. This is then related to $\iota^\tau_\infty(u, z)$ by

$$\iota^\tau_\infty(u, z) = \mp w_\tau(\beta_z). \tag{C.18}$$

176   *A Quick Survey of Siefring's Intersection Theory*

What Hutchings in [Hut02, Hut14] calls the "total" writhe $w_\tau(u) \in \mathbb{Z}$ of a simple holomorphic curve u is defined by adding up the writhes at all positive punctures, plus linking numbers for pairs of distinct punctures that have coinciding asymptotic orbits (up to multiplicity), and then subtracting all of the corresponding terms for the negative punctures. This produces

$$w_\tau(u) = -\iota_\infty^\tau(u);$$

thus the relative adjunction formula for a simple curve $u \colon \dot\Sigma \to \widehat{W}$ in ECH language takes the form

$$c_\tau(u) = \chi(u) + Q_\tau(u) + w_\tau(u) - 2\delta(u),$$

where

- $c_\tau(u) := c_1^\tau(u^*T\widehat{W})$ is the relative first Chern number.
- $\chi(u) := \chi(\dot\Sigma)$ is the Euler characteristic of the domain.
- $Q_\tau(u) := u \bullet_\tau u$ is the relative self-intersection number.
- $w_\tau(u)$ is the total writhe as explained above.
- $\delta(u)$ is the usual algebraic count of double points and critical points.

The reader can now check that this formula is equivalent to (4.7).

References

[Abb11] C. Abbas, *Holomorphic open book decompositions*, Duke Math. J. **158** (2011), no. 1, 29–82.

[ACH05] C. Abbas, K. Cieliebak, and H. Hofer, *The Weinstein conjecture for planar contact structures in dimension three*, Comment. Math. Helv. **80** (2005), no. 4, 771–793.

[AF03] R. A. Adams and J. J. F. Fournier, *Sobolev Spaces*, 2nd ed., Pure and Applied Mathematics (Amsterdam), vol. 140, Elsevier/Academic Press, Amsterdam, 2003.

[AEMS08] A. Akhmedov, J. B. Etnyre, T. E. Mark, and I. Smith, *A note on Stein fillings of contact manifolds*, Math. Res. Lett. **15** (2008), no. 6, 1127–1132.

[AD14] M. Audin and M. Damian, *Morse Theory and Floer Homology*, Universitext, Springer, London; EDP Sciences, Les Ulis, 2014. Translated from the 2010 French original by Reinie Erné.

[BEV12] K. L. Baker, J. B. Etnyre, and J. Van Horn-Morris, *Cabling, contact structures and mapping class monoids*, J. Differ. Geom. **90** (2012), no. 1, 1–80.

[BEM15] M. S. Borman, Y. Eliashberg, and E. Murphy, *Existence and classification of overtwisted contact structures in all dimensions*, Acta Math. **215** (2015), no. 2, 281–361.

[Bou02] F. Bourgeois, *A Morse–Bott approach to contact homology*, Ph.D. Thesis, Stanford University, 2002.

[Bou06] F. Bourgeois, *Contact homology and homotopy groups of the space of contact structures*, Math. Res. Lett. **13** (2006), no. 1, 71–85.

[BEH+03] F. Bourgeois, Y. Eliashberg, H. Hofer, K. Wysocki, and E. Zehnder, *Compactness results in symplectic field theory*, Geom. Topol. **7** (2003), 799–888.

[BM04] F. Bourgeois and K. Mohnke, *Coherent orientations in symplectic field theory*, Math. Z. **248** (2004), no. 1, 123–146.

[Bre93] G. E. Bredon, *Topology and Geometry*, Springer-Verlag, New York, 1993.

[CE12] K. Cieliebak and Y. Eliashberg, *From Stein to Weinstein and Back: Symplectic Geometry of Affine Complex Manifolds*, American Mathematical Society Colloquium Publications, vol. 59, American Mathematical Society, Providence, RI, 2012.

178 *References*

[CW] A. Cioba and C. Wendl, *Unknotted Reeb orbits and nicely embedded holomorphic curves*. Preprint arXiv:1609.01660, to appear in J. Symplect. Geom.

[CZ83] C. Conley and E. Zehnder, *An Index Theory for Periodic Solutions of a Hamiltonian System*, Geometric dynamics (Rio de Janeiro, 1981), Lecture Notes in Mathematics, vol. 1007, Springer, Berlin, 1983, pp. 132–145.

[CE] J. Conway and J. B. Etnyre, *Contact surgery and symplectic caps*. Preprint arXiv:1811.00387.

[Don99] S. K. Donaldson, *Lefschetz pencils on symplectic manifolds*, J. Differ. Geom. **53** (1999), no. 2, 205–236.

[Dra04] D. L. Dragnev, *Fredholm theory and transversality for noncompact pseudoholomorphic maps in symplectizations*, Comm. Pure Appl. Math. **57** (2004), no. 6, 726–763.

[Eli89] Y. Eliashberg, *Classification of overtwisted contact structures on 3-manifolds*, Invent. Math. **98** (1989), no. 3, 623–637.

[Eli90] Y. Eliashberg, *Filling by Holomorphic Discs and Its Applications*, Geometry of low-dimensional manifolds, 2 (Durham, 1989), London Mathematical Society Lecture Note Series, vol. 151, Cambridge University Press, Cambridge, 1990, pp. 45–67.

[Eli91] Y. Eliashberg, *On symplectic manifolds with some contact properties*, J. Differ. Geom. **33** (1991), no. 1, 233–238.

[EGH00] Y. Eliashberg, A. Givental, and H. Hofer, *Introduction to symplectic field theory*, Geom. Funct. Anal., Special Volume (2000), 560–673.

[EM] Y. Eliashberg and E. Murphy, *Making cobordisms symplectic*. Preprint arXiv:1504.06312.

[Etn98] J. B. Etnyre, *Symplectic convexity in low-dimensional topology*, Topology Appl. **88** (1998), no. 1–2, 3–25. Symplectic, contact and low-dimensional topology (Athens, GA, 1996).

[Etn04] J. B. Etnyre, *Planar open book decompositions and contact structures*, Int. Math. Res. Not. **79** (2004), 4255–4267.

[Etn06] J. B. Etnyre, *Lectures on Open Book Decompositions and Contact Structures*, Floer homology, gauge theory, and low-dimensional topology, Clay Mathematics Proceedings, vol. 5, American Mathematical Society, Providence, RI, 2006, pp. 103–141.

[EH01] J. B. Etnyre and K. Honda, *On the nonexistence of tight contact structures*, Ann. of Math. (2) **153** (2001), no. 3, 749–766.

[EH02] J. B. Etnyre and K. Honda, *On symplectic cobordisms*, Math. Ann. **323** (2002), no. 1, 31–39.

[Eva98] L. C. Evans, *Partial Differential Equations*, Graduate Studies in Mathematics, vol. 19, American Mathematical Society, Providence, RI, 1998.

[Gei08] H. Geiges, *An Introduction to Contact Topology*, Cambridge Studies in Advanced Mathematics, vol. 109, Cambridge University Press, Cambridge, 2008.

[Ghi05] P. Ghiggini, *Strongly fillable contact 3-manifolds without Stein fillings*, Geom. Topol. **9** (2005), 1677–1687.

[Gir02] E. Giroux, *Géométrie de contact: de la dimension trois vers les dimensions supérieures*, Proceedings of the International Congress of Mathematicians, vol. 2 (Beijing, 2002), 2002, pp. 405–414.

References

[GS99] R. E. Gompf and A. I. Stipsicz, *4-Manifolds and Kirby Calculus*, Graduate Studies in Mathematics, vol. 20, American Mathematical Society, Providence, RI, 1999.

[Gro85] M. Gromov, *Pseudoholomorphic curves in symplectic manifolds*, Invent. Math. **82** (1985), no. 2, 307–347.

[Hin00] R. Hind, *Holomorphic filling of* $\mathbf{R}P^3$, Commun. Contemp. Math. **2** (2000), no. 3, 349–363.

[Hin03] R. Hind, *Stein fillings of lens spaces*, Commun. Contemp. Math. **5** (2003), no. 6, 967–982.

[Hir94] M. W. Hirsch, *Differential Topology*, Springer-Verlag, New York, 1994.

[HLS97] H. Hofer, V. Lizan, and J.-C. Sikorav, *On genericity for holomorphic curves in four-dimensional almost-complex manifolds*, J. Geom. Anal. **7** (1997), no. 1, 149–159.

[HWZ95] H. Hofer, K. Wysocki, and E. Zehnder, *Properties of pseudo-holomorphic curves in symplectisations. II. Embedding controls and algebraic invariants*, Geom. Funct. Anal. **5** (1995), no. 2, 270–328.

[HWZ96a] H. Hofer, K. Wysocki, and E. Zehnder, *Properties of pseudoholomorphic curves in symplectisations. I. Asymptotics*, Ann. Inst. H. Poincaré Anal. Non Linéaire **13** (1996), no. 3, 337–379.

[HWZ96b] H. Hofer, K. Wysocki, and E. Zehnder, *Properties of Pseudoholomorphic Curves in Symplectisations. IV. Asymptotics with Degeneracies*, Contact and Symplectic Geometry (Cambridge, 1994), Cambridge University Press, Cambridge, 1996, pp. 78–117.

[HWZ99] H. Hofer, K. Wysocki, and E. Zehnder, *Properties of pseudoholomorphic curves in symplectizations. III. Fredholm theory*, Top. Nonlin. Anal. **13** (1999), 381–475.

[HZ94] H. Hofer and E. Zehnder, *Symplectic Invariants and Hamiltonian Dynamics*, Birkhäuser Verlag, Basel, 1994.

[Hry12] U. Hryniewicz, *Fast finite-energy planes in symplectizations and applications*, Trans. Amer. Math. Soc. **364** (2012), no. 4, 1859–1931.

[HMS15] U. Hryniewicz, A. Momin, and P. A. S. Salomão, *A Poincaré-Birkhoff theorem for tight Reeb flows on* S^3, Invent. Math. **199** (2015), no. 2, 333–422.

[Hut02] M. Hutchings, *An index inequality for embedded pseudoholomorphic curves in symplectizations*, J. Eur. Math. Soc. (JEMS) **4** (2002), no. 4, 313–361.

[Hut14] M. Hutchings, *Lecture Notes on Embedded Contact Homology*, Contact and Symplectic Topology, Bolyai Society Mathematical Studies, vol. 26, Springer, New York, 2014, pp. 389–484.

[IS99] S. Ivashkovich and V. Shevchishin, *Structure of the moduli space in a neighborhood of a cusp-curve and meromorphic hulls*, Invent. Math. **136** (1999), no. 3, 571–602.

[KL16] A. Kaloti and Y. Li, *Stein fillings of contact 3-manifolds obtained as Legendrian surgeries*, J. Symplect. Geom. **14** (2016), no. 1, 119–147.

[Kri98] M. Kriener, *Intersection formula for finite energy half cylinders*, Ph.D. Thesis, ETH Zürich, 1998.

[LM96] F. Lalonde and D. McDuff, *J-Curves and the Classification of Rational and Ruled Symplectic 4-Manifolds*, Contact and Symplectic

References

Geometry (Cambridge, 1994), Cambridge University Press, Cambridge, 1996, pp. 3–42.

[Laz] O. Lazarev, *Maximal contact and symplectic structures*. Preprint arXiv:1810.11728.

[LL01] E. H. Lieb and M. Loss, *Analysis*, 2nd ed., Graduate Studies in Mathematics, vol. 14, American Mathematical Society, Providence, RI, 2001.

[Lis98] P. Lisca, *Symplectic fillings and positive scalar curvature*, Geom. Topol. **2** (1998), 103–116.

[Lis08] P. Lisca, *On symplectic fillings of lens spaces*, Trans. Amer. Math. Soc. **360** (2008), no. 2, 765–799.

[LVWa] S. Lisi, J. Van Horn-Morris, and C. Wendl, *On symplectic fillings of spinal open book decompositions I: Geometric constructions*. Preprint arXiv:1810.12017.

[LVWb] S. Lisi, J. Van Horn-Morris, and C. Wendl, *On symplectic fillings of spinal open book decompositions II: Holomorphic curves and classification*. In preparation.

[Mar71] J. Martinet, *Formes de contact sur les variétés de dimension* 3, Proceedings of Liverpool Singularities Symposium, II (1969/1970), Springer, Berlin, 1971, pp. 142–163. Lecture Notes in Mathematics, vol. 209 (French).

[McD90] D. McDuff, *The structure of rational and ruled symplectic 4-manifolds*, J. Amer. Math. Soc. **3** (1990), no. 3, 679–712.

[McD94] D. McDuff, *Singularities and Positivity of Intersections of J-Holomorphic Curves*, Holomorphic curves in symplectic geometry, Progress in Mathematics, vol. 117, Birkhäuser, Basel, 1994, pp. 191–215. With an appendix by Gang Liu.

[MS12] D. McDuff and D. Salamon, *J-Holomorphic Curves and Symplectic Topology*, 2nd ed., American Mathematical Society Colloquium Publications, vol. 52, American Mathematical Society, Providence, RI, 2012.

[MS17] D. McDuff and D. Salamon, *Introduction to Symplectic Topology*, 3rd ed., Oxford University Press, Oxford, 2017.

[MW95] M. J. Micallef and B. White, *The structure of branch points in minimal surfaces and in pseudoholomorphic curves*, Ann. Math. (2) **141** (1995), no. 1, 35–85.

[Mil97] J. W. Milnor, *Topology from the Differentiable Viewpoint*, Princeton Landmarks in Mathematics, Princeton University Press, Princeton, NJ, 1997. Based on notes by David W. Weaver; revised reprint of the 1965 original.

[Mor03] E. Mora, *Pseudoholomorphic cylinders in symplectisations*, Ph.D. Thesis, New York University, 2003.

[NW11] K. Niederkrüger and C. Wendl, *Weak symplectic fillings and holomorphic curves*, Ann. Sci. École Norm. Sup. (4) **44** (2011), no. 5, 801–853.

[OS04a] B. Ozbagci and A. I. Stipsicz, *Surgery on Contact 3-Manifolds and Stein Surfaces*, Bolyai Society Mathematical Studies, vol. 13, Springer-Verlag, Berlin, 2004.

[OS04b] B. Ozbagci and A. I. Stipsicz, *Contact 3-manifolds with infinitely many Stein fillings*, Proc. Amer. Math. Soc. **132** (2004), no. 5, 1549–1558.

[PV10] O. Plamenevskaya and J. Van Horn-Morris, *Planar open books, monodromy factorizations and Stein fillings*, Geom. Topol. **14** (2010), 2077–2101.

References

[Sal99] D. Salamon, *Lectures on Floer Homology*, Symplectic Geometry and Topology (Park City, UT, 1997), IAS/Park City Mathematics Series, vol. 7, American Mathematical Society, Providence, RI, 1999, pp. 143–229.

[SZ92] D. Salamon and E. Zehnder, *Morse theory for periodic solutions of Hamiltonian systems and the Maslov index*, Comm. Pure Appl. Math. **45** (1992), no. 10, 1303–1360.

[Sar42] A. Sard, *The measure of the critical values of differentiable maps*, Bull. Amer. Math. Soc. **48** (1942), 883–890.

[Sch93] M. Schwarz, *Morse Homology*, Progress in Mathematics, vol. 111, Birkhäuser Verlag, Basel, 1993.

[Sie05] R. Siefring, *Intersection theory of finite energy surfaces*, Ph.D. Thesis, New York University, 2005.

[Sie08] R. Siefring, *Relative asymptotic behavior of pseudoholomorphic half-cylinders*, Comm. Pure Appl. Math. **61** (2008), no. 12, 1631–1684.

[Sie11] R. Siefring, *Intersection theory of punctured pseudoholomorphic curves*, Geom. Topol. **15** (2011), 2351–2457.

[SW] R. Siefring and C. Wendl, *Pseudoholomorphic curves, intersections and Morse–Bott asymptotics*. In preparation.

[Sik94] J.-C. Sikorav, *Some Properties of Holomorphic Curves in almost Complex Manifolds*, Holomorphic Curves in Symplectic Geometry, Progress in Mathematics, vol. 117, Birkhäuser, Basel, 1994, pp. 165–189.

[Sik97] J.-C. Sikorav, *Singularities of J-holomorphic curves*, Math. Z. **226** (1997), no. 3, 359–373.

[Smi01] I. Smith, *Torus fibrations on symplectic four-manifolds*, Turkish J. Math. **25** (2001), no. 1, 69–95.

[Thu76] W. P. Thurston, *Some simple examples of symplectic manifolds*, Proc. Amer. Math. Soc. **55** (1976), no. 2, 467–468.

[TW75] W. P. Thurston and H. E. Winkelnkemper, *On the existence of contact forms*, Proc. Amer. Math. Soc. **52** (1975), 345–347.

[Wan12] A. Wand, *Mapping class group relations, Stein fillings, and planar open book decompositions*, J. Topol. **5** (2012), no. 1, 1–14.

[Wan15] A. Wand, *Factorizations of diffeomorphisms of compact surfaces with boundary*, Geom. Topol. **19** (2015), no. 5, 2407–2464.

[Wen10a] C. Wendl, *Automatic transversality and orbifolds of punctured holomorphic curves in dimension four*, Comment. Math. Helv. **85** (2010), no. 2, 347–407.

[Wen10b] C. Wendl, *Strongly fillable contact manifolds and J-holomorphic foliations*, Duke Math. J. **151** (2010), no. 3, 337–384.

[Wen10c] C. Wendl, *Open book decompositions and stable Hamiltonian structures*, Expos. Math. **28** (2010), no. 2, 187–199.

[Wena] C. Wendl, *Lectures on Holomorphic Curves in Symplectic and Contact Geometry*. Preprint arXiv:1011.1690.

[Wenb] C. Wendl, *Lectures on Symplectic Field Theory*. Preprint arXiv:1612.01009, to appear in EMS Series of Lectures in Mathematics.

[Wen18] C. Wendl, *Holomorphic Curves in Low Dimensions: From Symplectic Ruled Surfaces to Planar Contact Manifolds*, Lecture Notes in Mathematics, vol. 2216, Springer-Verlag, Berlin, 2018.

Index

adjunction formula
  for closed holomorphic curves, 31
  for punctured holomorphic curves, 65–76, 163
  relative, 72, 176
almost complex manifold with cylindrical ends, 40, 160
almost complex structure, 14
  compatible with a contact form, 40
  compatible with a stable Hamiltonian structure, 160
  compatible with a symplectic form, 14
  integrable, 13, 18
  tamed by a symplectic form, 14
arithmetic genus, 97, 105
asymptotic contribution
  to the $*$-pairing, 70
  to the singularity index, 75
asymptotic defect, 56
asymptotic eigenfunction, 51
  extremal winding of, 70
  relative, 52
asymptotic operator, 48, 159
asymptotic positivity of intersections, 70
asymptotic Reeb orbit, 41
asymptotic representative of a punctured holomorphic curve, 51
asymptotic trivializations, 57–58, 159
asymptotic winding of a section, 56
asymptotically cylindrical map, 41, 160
  relative homology class of, 42
automatic transversality
  for closed holomorphic curves, 96
  for punctured holomorphic curves, 104
automorphism group

of a closed holomorphic curve, 95
of a punctured holomorphic curve, 103

Baire category theorem, 96
Baire set, 96
biholomorphic, 19
binding of an open book, 77
blowup
  of a complex manifold, 16
  of a symplectic manifold, 16
bordered Lefschetz fibration, 79
  allowable, 83
  of a symplectic filling, 84
braid
  linking number, 175
  writhe, 175
breaking contribution, 172
breaking orbits, 105, 168
  can hide intersections, 171–172

Cauchy–Riemann type operator, 109
  complex linear, 109, 111, 120
  of class C^m, 110
circle compactification of a punctured Riemann surface, 41
comeager, 96
compactification of moduli spaces, 20, 97, 105, 168
compatible almost complex structure, 14, 40, 160
completion of a symplectic cobordism, 39
complex blowup, 16
complex manifold, 13
complex structure
  on a manifold, 13
  on a vector bundle, 14

Index

183

concatenation of almost complex manifolds with cylindrical ends, 168
concave boundary, 35
Conley–Zehnder index, 54
connected components of a nodal holomorphic curve, 168
contact action functional, 47
contact form, 35
contact manifold, 35
  planar, 83
contact structure, 35
  on $S^1 \times S^2$, 38
  on S^3, 37
  on lens spaces, 38
  planar, 83
  supported by an open book, 78
contact-type boundary, 35
convex boundary, 35
covering multiplicity
  of a Reeb orbit, 64
  of an asymptotic eigenfunction, 65, 164
critical order, 127, 128
critical points of a holomorphic curve, 29, 95
cylindrical coordinates on a punctured Riemann surface, 40, 174
cylindrical ends
  of a Riemann surface, 40
  of an almost complex manifold, 40, 160

double points
  hidden at infinity, 65
  of a holomorphic curve, 29

elliptic orbit, 159
energy
  of a closed holomorphic curve, 97
  of a punctured holomorphic curve, 104
even Reeb orbit, 54; *see also* parity of a Reeb orbit
exceptional sphere, 16
extension of a holomorphic building, 169
extremal winding, 70
extremal winding numbers of a Reeb orbit, 54, 159

fibration
  Lefschetz, 13
  symplectic, 12
Fredholm regular, 95–96, 103–104

generalized tangent-normal splitting, 128
generic, 96

Giroux form, 78
Gromov's compactness theorem, 98–100

hidden at infinity
  double points, 65
  intersections, 63, 70
  zeroes of a section, 56
Hofer energy, 104
holomorphic building
  breakings orbits of, 168
  extension of, 169
  in a concatenation, 168
  levels, 105, 168
  stable, 105, 169
holomorphic curve, 18
  asymptotic eigenfunction at a puncture, 51
  asymptotic representative at a puncture, 51
  asymptotically cylindrical, 41, 160
  critical order of a critical point, 127, 128
  critical points of, 29, 95
  decay rate at a puncture, 51
  double points of, 29
  energy of, 97, 104
  generalized normal bundle of, 128
  immersed points of, 29, 95
  index of, 59, 94, 102
  local singularity index at a point, 30, 152–157
  multiply covered, 19, 94, 103, 151–152
  nodal, 20–21, 97, 168
  normal Chern number of, 30, 58, 163
  regular, 95–96, 103–104
  relative asymptotic eigenfunction of two punctures, 52
  relative decay rate of two punctures, 52
  simple, 19, 94–95, 103
  singularity index of, 30, 164
  spectral covering number of, 65, 164
  tangent space at a critical point, 127–130
holomorphic open book, 86
holomorphic vector bundle, 111, 120
horizontal boundary of a Lefschetz fibration, 79
hyperbolic orbit, 159

immersed points of a holomorphic curve, 29, 95
index
  of a closed holomorphic curve, 94
  of a punctured holomorphic curve, 59, 102
injectivity modulus, 153

184   *Index*

intersection number
  asymptotic contribution to, 70
  homological, 26
  local, 26
  of asymptotically cylindrical maps
    (∗-pairing), 63, 70
intersections
  hidden at infinity, 63, 70
  hidden between levels of a building,
    171–172
  positivity of, 28, 148–151
irreducible components of a Lefschetz
  singular fiber, 13

J-holomorphic curve, *see* holomorphic curve

Lefschetz fibration, 13
  allowable, 83
  as filling of an open book, 84
  bordered, 79
  horizontal boundary of, 79
  irreducible components of a singular fiber,
    13
  monodromy, 79–81
  of a symplectic filling, 84
  regular fiber of, 13
  singular fiber of, 13, 97, 105
  symplectic, 15
  vanishing cycle, 80
  vertical boundary of, 79
linking number of two braids, 175
Liouville form, 35
Liouville vector field, 35
local intersection index, 26
local singularity index, 30, 152–157

Micallef–White theorem, 126, 127
minimal symplectic manifold, 17
moduli space
  compactification of, 20, 97, 105, 168
  of closed holomorphic curves, 19
  virtual dimension of, 94, 95, 103
monodromy, 79–81
Morse function, 45
Morse–Bott Reeb orbits, 158
multiply covered holomorphic curve, 19, 94,
  103, 151–152
multiply covered Reeb orbit, 54

negative hyperbolic orbit, 160
negative punctures of a holomorphic curve, 40
nodal holomorphic curve, 20–21, 97

nodal points, 168
nodes of a holomorphic curve, 20, 97, 168
nondegenerate Reeb orbit, 49
  Conley–Zehnder index of, 54
normal Chern number
  as a count of zeroes, 58
  of a closed holomorphic curve, 30
  of a punctured holomorphic curve, 58, 163

odd Reeb orbit, 54, 159
open book decomposition, 77
  binding of, 77
  filled by a Lefschetz fibration, 84
  holomorphic, 86
  monodromy of, 79–81
  pages of, 78
  planar, 83
  supporting a contact structure, 78
orbit cylinder, 40, 76, 88, 106

pages of an open book, 78
parity of a Reeb orbit, 54, 159
planar contact manifold, 83
planar open book, 83
positive punctures of a holomorphic curve, 40
positivity of intersections, 28, 148–151
  asymptotic, 70
pseudoholomorphic curve, *see* holomorphic
  curve

Reeb orbit
  asymptotic, 41
  asymptotic operator of, 48, 159
  breaking, 105, 168
  Conley–Zehnder index of, 54
  covering multiplicity of, 64
  elliptic, 159
  even/odd, 54
  extremal winding numbers of, 54, 159
  hyperbolic, 159
  Morse–Bott, 158
  multiply covered, 54
  negative hyperbolic, 160
  nondegenerate, 49
  parity of, 54, 159
  simply covered, 54
  spectral covering number of, 65, 164
Reeb vector field
  of a contact form, 39
  of a stable Hamiltonian structure, 158
regular fiber of a Lefschetz fibration, 13
relative adjunction formula, 72, 176

Index

relative asymptotic eigenfunction, 52
relative decay rate, 52
relative first Chern number, 57, 163
relative homology class, 42
relative intersection number, 66, 161
Riemann surface, 18
  circle compactification of, 41
  punctured, 40
ruled surface, 12

second category, 96
self-intersection number
  homological, 16, 23, 30
  of asymptotically cylindrical maps, 65, 75, 88, 163
  relative, 72, 176
SFT compactness theorem, 104–107
similarity principle, 24–25, 120–126
  asymptotic analogue, 50
simple holomorphic curve, 19, 94–95, 103
  is locally injective, 151–152
simply covered Reeb orbit, 54
singular fiber of a Lefschetz fibration, 13, 97, 105
singularity index of a simple holomorphic curve, 30, 164
  asymptotic contribution to, 75
Sobolev spaces, 49, 110, 111
somewhere injective, *see* simple holomorphic curve
spectral covering number, 65, 164
stability
  of a holomorphic building, 106
  of a nodal holomorphic curve, 97
stable Hamiltonian structure, 158
standard contact structure
  on $S^1 \times S^2$, 38
  on S^3, 37
  on lens spaces, 38
Stein filling, 83, 85
symplectic blowup, 16
symplectic cap, 36
symplectic cobordism, 36
symplectic completion, 39
symplectic deformation equivalence, 37

symplectic fibration, 12
symplectic filling, 36
  Lefschetz fibrations on, 84
  Stein, 83, 85
  weak, 85
  Weinstein, 83, 85
symplectic form, 11
  supported by a bordered Lefschetz fibration, 82–83
  supported by a Lefschetz fibration, 15
symplectic Lefschetz fibration, 15
symplectic manifold, 11
  minimal, 17
symplectic ruled surface, 12, 18
symplectic structure
  on a manifold, 11
  on a vector bundle, 14
symplectic submanifold, 11
symplectically immersed, 29, 153, 154
symplectization of a contact manifold, 39
symplectomorphism, 12

tame almost complex structure, 14
tangent space of a holomorphic curve, 127–130
transversality
  automatic, 96, 104
  for holomorphic curves, *see* Fredholm regular
  for intersections, 26
transverse, 26
trivial cylinder, 40, 76, 88, 106

unique continuation, 28, 95

vanishing cycle, 80
vertical boundary of a Lefschetz fibration, 79
virtual dimension, 94, 95, 103

weak symplectic filling, 85
Weinstein filling, 83, 85
writhe of a braid, 175

zeroes of a section
  hidden at infinity, 56
  positivity of, 24, 128